THE
ACCIDENTAL
BOTANIST

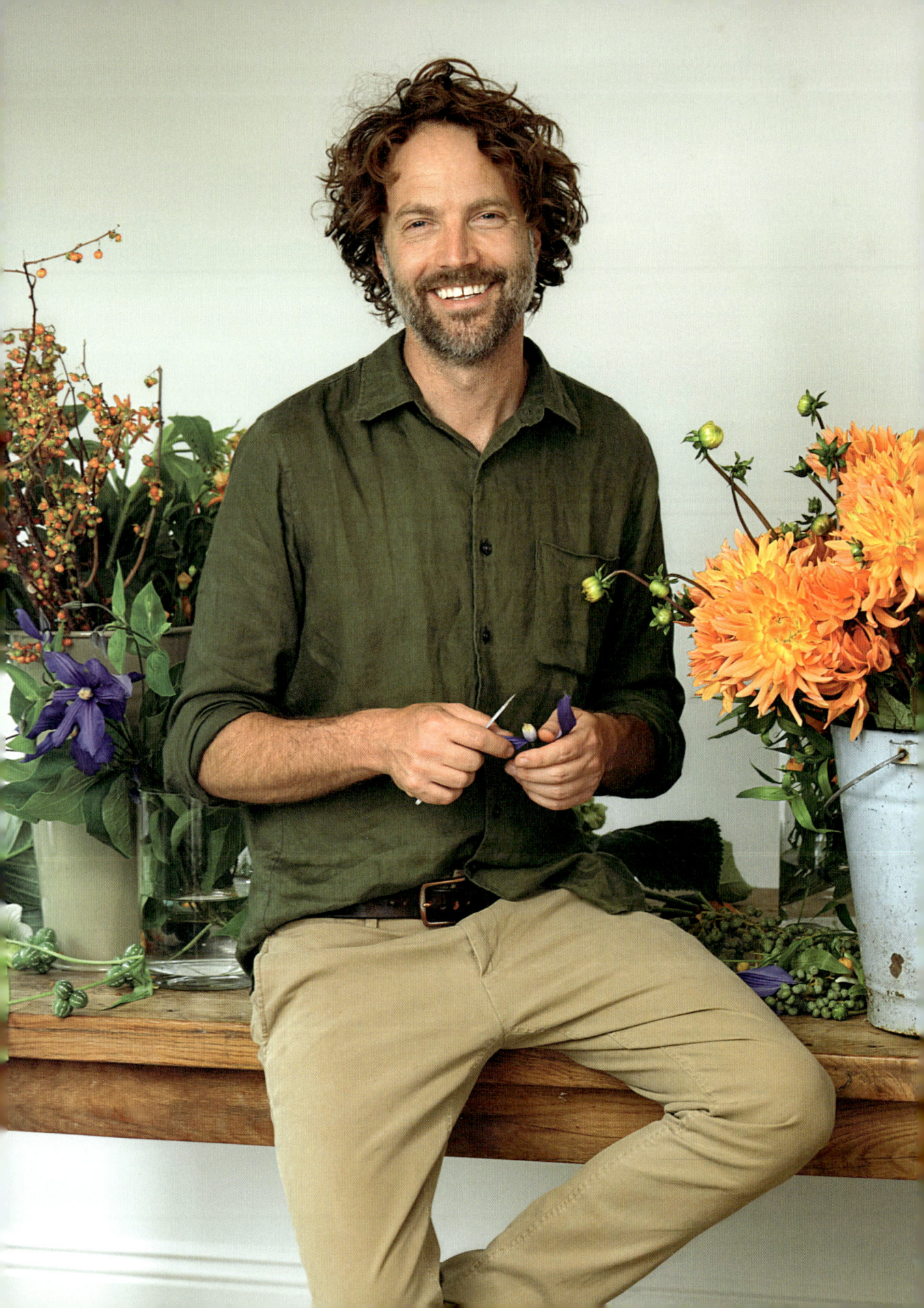

ROBBIE HONEY .

THE ACCIDENTAL BOTANIST

The Structure of Plants Revealed

CLEARVIEW

FOR A.H.

First published in the UK in 2018 by Clearview Books,
22 Clarendon Gardens, London W9 1AZ

Compilation © 2018 Clearview Books, London
Text and images © 2018 Robbie Honey

Second printing 2018

ISBN: 978-1908337-443

www.clearviewbooks.com

Publisher: Catharine Snow
Photography: Robbie Honey and Katrina Lawson Johnston
Design: Bernard Higton
Editor: Jemima Dunne
Production: Simonne Waud

A CIP record of this book is available from
The British Library.

Colour repro by XY Digital Ltd, London
Printed in Croatia by Zrinski D.D.

CONTENTS

INTRODUCTION 6

1

RED 12

2

ORANGE 30

3

YELLOW 44

4

BLUE AND PURPLE 64

5

PINK 94

6

WHITE 116

GLOSSARY 156

INDEX 157

ACKNOWLEDGEMENTS 160

INTRODUCTION

My family has been on the African continent for five generations, four of those in Zimbabwe. I consider myself entirely Zimbabwean and am proud to be an African, albeit a white one.

My father's grandparents, William and Catherine Honey, arrived in Rhodesia in 1893. Their youngest son Robert (always known as Bert), my grandfather and my namesake, went on to study medicine in Edinburgh, Scotland. His wife, my grandmother, Molly Cathcart, grew up in Salisbury in an Italianate villa, Chinanga, designed by her architect father, D'arcy Cathcart. They married in Cape Town, South Africa, and returned to Salisbury where they lived in a pretty Dutch-gabled house, and my grandfather, a urologist, worked for much of his life at the nearby St Anne's hospital, run by nuns of the Little Company of Mary. He was later awarded a *Benemerenti* medal for services to the hospital by Pope Paul VI, and in 1964 received an OBE for his work on the treatment of Bilharziasis (caused by a water-born parasite) and for his public services.

After a few years, with three boys and a fourth child on the way, they purchased 20 acres of land in the emerging suburb of Rietfontein on the outskirts of Salisbury, and built Maduma, a large, shuttered villa. My grandfather carved out a six-acre terraced garden that surrounded the house. His design was simple. He planted armies of blue and white agapanthus that 'marched' along terraces in ranks, and hedges of hydrangeas, in all shades of pink and blue. The long drive to the house was lined with Naartjie trees (a *hesperidium* much like a satsuma). There was elegance in the simplicity of his approach and his style and it has most definitely influenced my preference for arrangements made up of one type of flower, rather than a multitude of different ones.

My mother's grandfather Neophytor Hajii Argiriou, arrived in Africa by boat from Cyprus in the early 20th century. The boat docked in Beira, Mozambique, and he headed for Rhodesia. Family legend has it that on arrival at the border post, he was asked for his passport. Speaking no English and without a passport, he just kept repeating the word 'passport' back to the officials in his strong Greek accent. The officials presumed this to be his name, so, Neophytor Passaportis crossed the border to his new life bearing a new name. At the time, those of Greek ethnicity were considered 'non-white' and were not only required to carry an identification document called a 'stoopa', but also were not permitted to travel by rail. So Neophytor walked towards Salisbury, and a month (and 180 miles) later, he arrived at the town of Gadzema, where he found a small Greek community. Here, he met and married Fotini, also a Cypriot. They had a daughter and three sons. My mother's father, their middle son, was born in 1922 and christened Theofani Lefkios Passaportis, later known as Theo. A talented sportsman Theo played rugby and cricket for Rhodesia, and commanded the 2nd battalion, King's African Rifles in Nyasaland. When in 1965 Ian Smith, Rhodesia's then president, announced a unilateral declaration of independence (UDI) from Britain, Theo was the military attaché in London and he, and his cadets at Sandhurst, were given 24 hours to pack up and leave the country. My grandmother, Ethelwyn Brent, was born in King Williamstown. She was a great beauty and a good match for my grandfather – and I adored her. They had two daughters and a son. My mother was the eldest and was christened Antoinette (known as Toni) Passaportis.

My parents first met in their teens, and two and a half years later on returning from national service my

Opposite: I have been captivated by flowers since I was a young boy. This is me aged five, *circa* 1980, in Lincolnshire, UK.

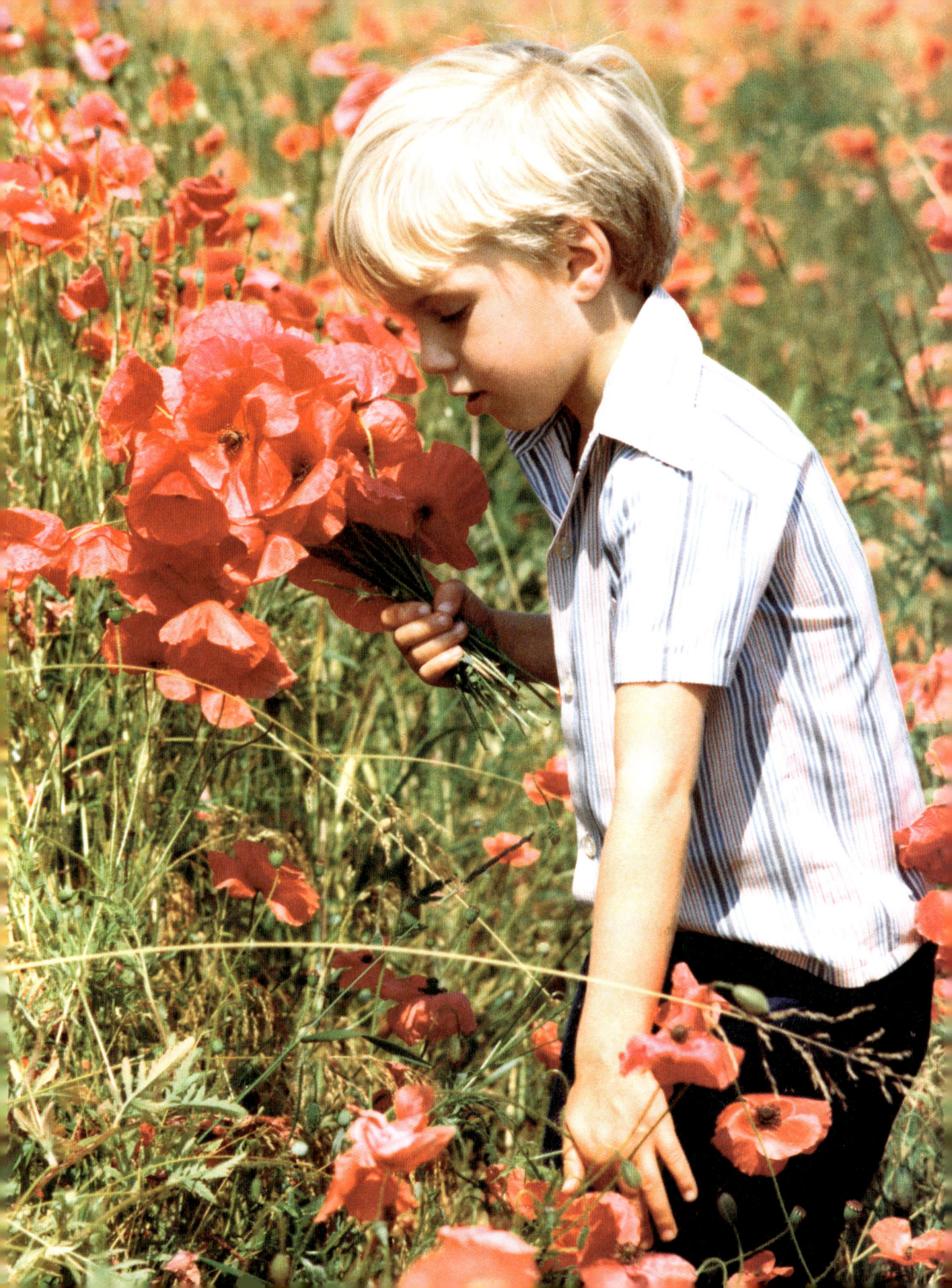

father proposed. They announced their engagement at my mother's 21st birthday party. A year later in 1969 – both aged 22 years – they married in the Salisbury's Anglican cathedral. My father's parents gave them an acre of the land at Maduma as wedding present, where they built their own house, Chinakwaremba. The two houses were separated only by a hedge of red hibiscus. My brother Richard was born in 1972, I followed in 1975. We had the freedom of both properties and loved spending time with our grandfather in his vegetable garden, which was laid out and managed in military precision. My grandmother grew flowers to cut for arranging – she had roses, gardenias, chinkirechees, and tuberoses. Adjacent to the vegetable garden were lychees, mangos, guavas, and an acre of bananas. There was a chicken run, which the hens shared with Mr and Mrs Lubbytum, a pair of ancient tortoises that my father and his brothers had had since they were boys. Alongside the chickens, was an empty enclosure where once, Elvis, a crocodile, had lived. Elvis escaped, and it always amused me to imagine him appearing unexpectedly in somebody else's swimming pool.

In 1979, after 15 years of fighting between the white Rhodesian government and the black majority, Ian Smith acquiesced to Robert Mugabe and Joshua Nkomo. In1980, independence was declared, Rhodesia became Zimbabwe, and Robert Mugabe was sworn in as president. In the wake of the change, whites left the country in droves, but my parents chose to stay. That same year my grandfather Bert, having performed a six-hour operation during the day, died in his sleep at the age of 76. My grandmother Molly moved into our house, and we went to live in Maduma.

For us postcolonial life in Zimbabwe remained much the same as before. We had a cook, housekeeper, gardeners, and a nanny, Stella, known as Lala, who looked after my brother and me. She kept me enchanted with her vivid imagination, a sense of wonder and a capacity to make the simplest excursion an adventure. She went on to look after my grandmother, and 46 years later, now officially retired, she is still part of the family and a great-grandmother herself. In between tending her fields, and ironing and flower arranging in the house, she gets up to mischief with my youngest niece, Jessica. Lala's surname is Dirorimwe, which means 'One Baboon' in Shona – Mrs One Baboon you are a marvel!

One of the many joys of an African childhood was that I was able to spend the majority of my waking hours barefoot and outdoors. I didn't particularly take to school and was happier gardening or climbing trees with my cat, Marmaduke. Over the years we looked after a menagerie of orphaned animals. We had a couple of vervet monkeys, Roger and Louis, who were always up no good, and we bottle-fed a shy pretty red duiker fawn, named Bambi.

At the age of 11 years, I became the youngest member of The Orchid Society of Zimbabwe. I was an avid collector, focussing on indigenous tree orchids. My father built me an orchid house complete with irrigation and mister sprays. I would set off into the bush climbing trees to collect them without realising that I required a permit to do so. Family friends still tell the story of a camping trip, when over the week I slowly palmed off the entire contents of my backpack in order to fill it with plants. My Uncle Patrick also had a keen interest in orchids and owned a property up in the Vumba (meaning misty mountains) that, apart from a small clearing with a stone bothy and a vegetable garden, was covered in impenetrable Montane cloud forest, where orchids abound. On weekends, we would head out into the forest armed with pangas to clear paths and gather orchids.

My family remain in Zimbabwe and, as 'townies', we did not lose our property in the land acquisition programme. In 2010, my brother Richard, his wife Emma, and their three children, Molly, Rory, and Jessica came back to Zimbabwe from London. They moved into the big house and my parents back into Chinakwaremba. It is a very happy setup, and I love seeing the children run barefoot in the vegetable garden and picking mulberry leaves for silkworms, much as I did as a boy.

Life in Zimbabwe is not easy. Much of the population lives below the poverty line and swathes of farmland lies fallow, so Zimbabwe must import most of its staple diet – maize – ironically, often from ex-Zimbabweans farming in other African countries. There is 90 percent unemployment, municipal water rarely flows in the cities, and electricity is sporadic. After 37 years in office, Robert Mugabe recently stepped down following an unofficial coup. Harare erupted into a huge street party to celebrate the hope of a different future. Can the new president, Emmerson Mnangagwa, a former henchman of

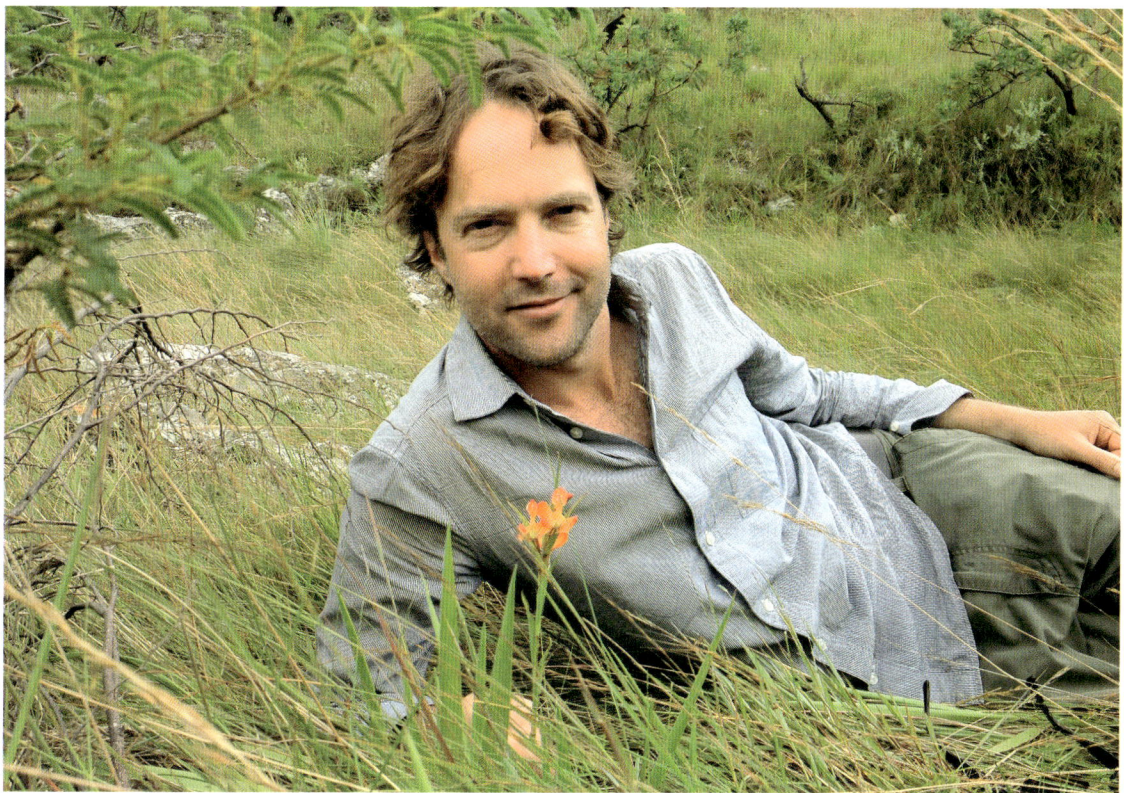

Every January I head to Zimbabwe's Eastern Highlands on a solo camping trip and spend my days hiking in search of indigenous flora.

Mugabe, bring about the much-needed change in our country? There is always hope.

I left school at the earliest opportunity. Flowers were always my passion, but floristry wasn't on the list of careers suggested, so I settled on becoming a rose farmer. I took an internship at a tissue culture and microbiology lab in Harare, where I learned the basics of micro-propagation. Then, at the age of 17, I went to Sparsholt college in Hampshire, UK, to study horticulture. Botany formed a small part of the course and I loved and excelled at it, but much of main syllabus didn't interest me. After a year I undertook some work experience on flower farms in Holland and Kenya, where I discovered that flower farming was much like any other form. My romantic illusions of being a rose farmer shattered, I knew that I wanted to work with plants, but wanted to do something creative. I returned to Zimbabwe to work for The Hamilton Kings, an interior decorator and landscape gardening duo. A year later I started

my own landscaping company, however, 1995 saw a year of severe drought, which left the country with little water for gardening and no work for a garden designer. So, I set off to study interior design in Cape Town. There I discovered I really did have an eye for colour, texture, and design, however, the course also highlighted my difficulty with technical drawing. Hampered by dyscalculia, my struggle with numbers made drawing perspectives challenging, so I followed this with a year studying photography at art school.

At the age of 23, with experience in horticulture, garden and interior design, and photography under my belt, I moved to London. Leafing through *Tatler*'s guide to 'Who's who' in the UK party industry I thought it looked like an exciting career prospect. London florist, Ming Veevers Carter, agreed to take me on as an apprentice. Her approach to flowers was bold, contemporary, and exciting and she threw me into the deep end. I started work at 3am and often found myself working seven days a week.

I loved the work, and Ming had a significant influence on my emerging style. A year later, the 3am starts became less appealing and, eager to see more of the industry, I started working freelance for other florists and began to pick up my own work. With the assistance of fellow Zimbabwean, Kimball Groves, I set up my eponymous company with all the blind confidence of youth, and no real idea of what it would take to build and manage a business. I began by arranging flowers out of the kitchen of the flat I shared with friends. At a time when most florists were using a myriad of varieties of flowers and greenery in one bunch, I chose to use only one type of flower, spiralled into concise bouquets. Each bunch was in a grey box tied with a monogrammed ribbon.

Some amazing opportunities followed. Stephanie Hoppen asked me to arrange flowers first for her book *White on White*, then her next book *Seeing Red*. Justin Van Breda, a friend from Cape Town, who was working for London decorator Nicky Haslam, hired me to arrange flowers for Haslam's table at a charity gala. The design included a mauve silk parasol set above a silk-swathed table. Inside the parasol, I fixed branches of copper beech adorned with twinkling lights and on the table below I placed autumnal hydrangea studded with mauve roses and figs. Haslam loved it. In 2001, I was asked to do the flowers for a Hermès fashion show in London's Soho. I was soon delivering weekly flowers to their London stores. One of the loveliest pieces I created for them – in collaboration with another designer Lulu Goodman – was a flower curtain for the entrance of their Sloane Street store. Made up of 2,000 Flame lilies, the 'curtain' was swept back and secured with a boss of *Rouge noir* roses, tasselled with cerise *amaranthus*. Lulu and I made an excellent team. Her 30-year experience in the industry, coupled with project management skills and an ability to drive a spreadsheet, allowed me to spend more time designing. This was the beginning of an 11-year partnership during which we arranged flowers for countless events, weddings, and parties. Together we designed a garden for Hermès, inspired by their scent, *Un Jardin en Méditerranée*, to celebrate the Chelsea Flower Show. On the strength of this Dior asked me to create gardens at their London head office. Soon I was arranging flowers for their press events and delivering their weekly bouquets, too.

Subsequent fragrance launches included flowers for Marc Jacobs, Balenciaga, Westwood, Tom Ford, Vera Wang, and Calvin Klein. I presented a Channel 5 show called *Nice house, shame about the garden* in which with the aid of a landscaper, an Irish builder, and a glamour model called Emma, derelict gardens were transformed into verdant oases.

When my own company was listed in the *Tatler*'s little black book of Who's Who in the events industry, I evaluated my work–life balance and realised that I didn't see myself working in the industry ten years hence. Within three months, I had closed my studio, sold my lease and props, and passed my client list on to Lulu. I then took a year's sabbatical.

In 2012, I started teaching floristry at FlowerSchool, New York. A year later I was invited by Cohim, a fashion college in Beijing, China to teach a series of floristry courses. I now spend 12 weeks a year at Cohim's FlowerSchool, Beijing. I teach, lecture, and demonstrate around the world working with many inspiring florists. I also continue to work as a floral stylist.

I have an acute sense of smell, and an interest in the scents of the natural world as well as those in the world of fine perfumery. Back in 1997, I met Victoria Spencer on a flight from South Africa to London. She was later to give me my first scented candle – a *Diptyque Figuier*. I could not believe how accurately the aroma of a hot fig tree was captured in the candle's wax; it instantly triggered memories of childhood holidays in the south of France. I knew then that one day I wanted to play the fragrance game, too. Years later, I set up a scented candle business with a friend Will Johnson. Combining my botanical knowledge with his ability to disseminate a scent into layers, we set about creating believable reconstructions of the scents of flowers for a capsule collection of candles under my name. In 2015, we successfully launched them – Lily of the valley, Jasmine, Casablanca lily, and Tuberose (the latter inspired by the tuberoses grown by my grandmother). Although I am not a nose, nor technically a perfumier, I have olfactory vision, knowledge, and a passion for scent. I now consult with other brands on botanical fragrances, too.

So, this book brings together the many facets that make me who I am. Several years ago, I started sketching for ten minutes a day. My first drawing was of a ginger lily flower (*Hedychium gardneranium*).

I absent-mindedly took it apart and drew the sum of its parts alongside my sketch of the flower. That was the start of the process, I believe, which has led to the collection of photographs for this book.

All my photographs are taken on an iPhone using natural light, and initially shot for Instagram – the shape initially dictated by the platform's then 'square-only format'. When photographing a specimen, I make a quick initial layout; the result is usually quite crowded. However, this starting point allows me to familiarise myself with the plant material and how it fits together. I have two essential bits of equipment, a ruler to keep arrangements uniform, and tweezers to re-arrange the delicate elements. It takes about an hour and roughly 50 frames to create a well-resolved composition. Once achieved, my initial specimens have generally wilted, so I collect fresh material to deconstruct, and arrange the elements on a clean sheet of textured pastel grey paper for the final frame. It is important to note that my deconstructions are not academic or complete representations of each specimen, nor are they meant to be. I capture what I perceive are the most beautiful attributes of a plant or flower.

The book's chapters are arranged by colour, which will give you some interesting and unexpected combinations and scales. I have provided botanical information for every plant, and the elements in the photographs are listed in a clockwise order starting from the left. All the species shown have their own narrative, as each one reminds me of places, events, or people I have met over the years of my collecting and studying plants.

It has given me great joy capturing the pictures, and I am delighted to be able to share them with you.

Robbie Honey, Harare 2018.

Studying and photographing the plants and flowers in this book has given me immense pleasure.

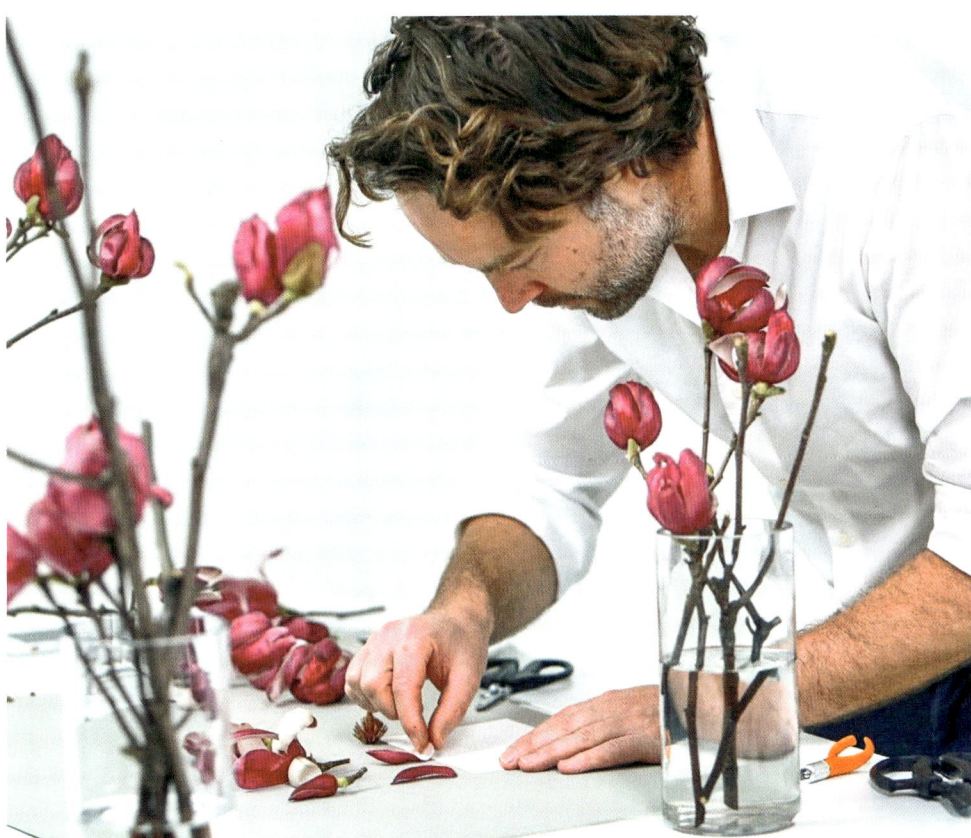

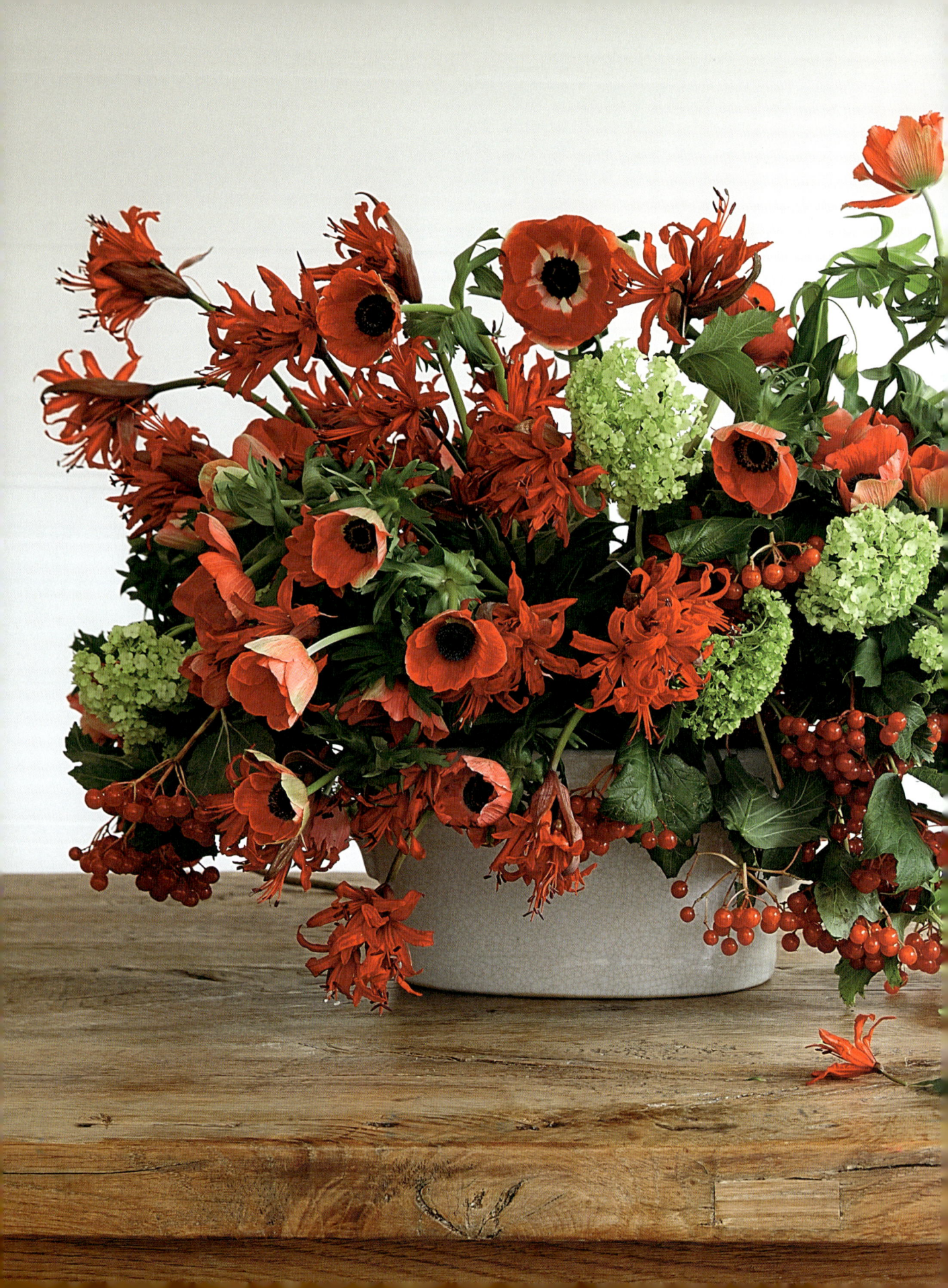

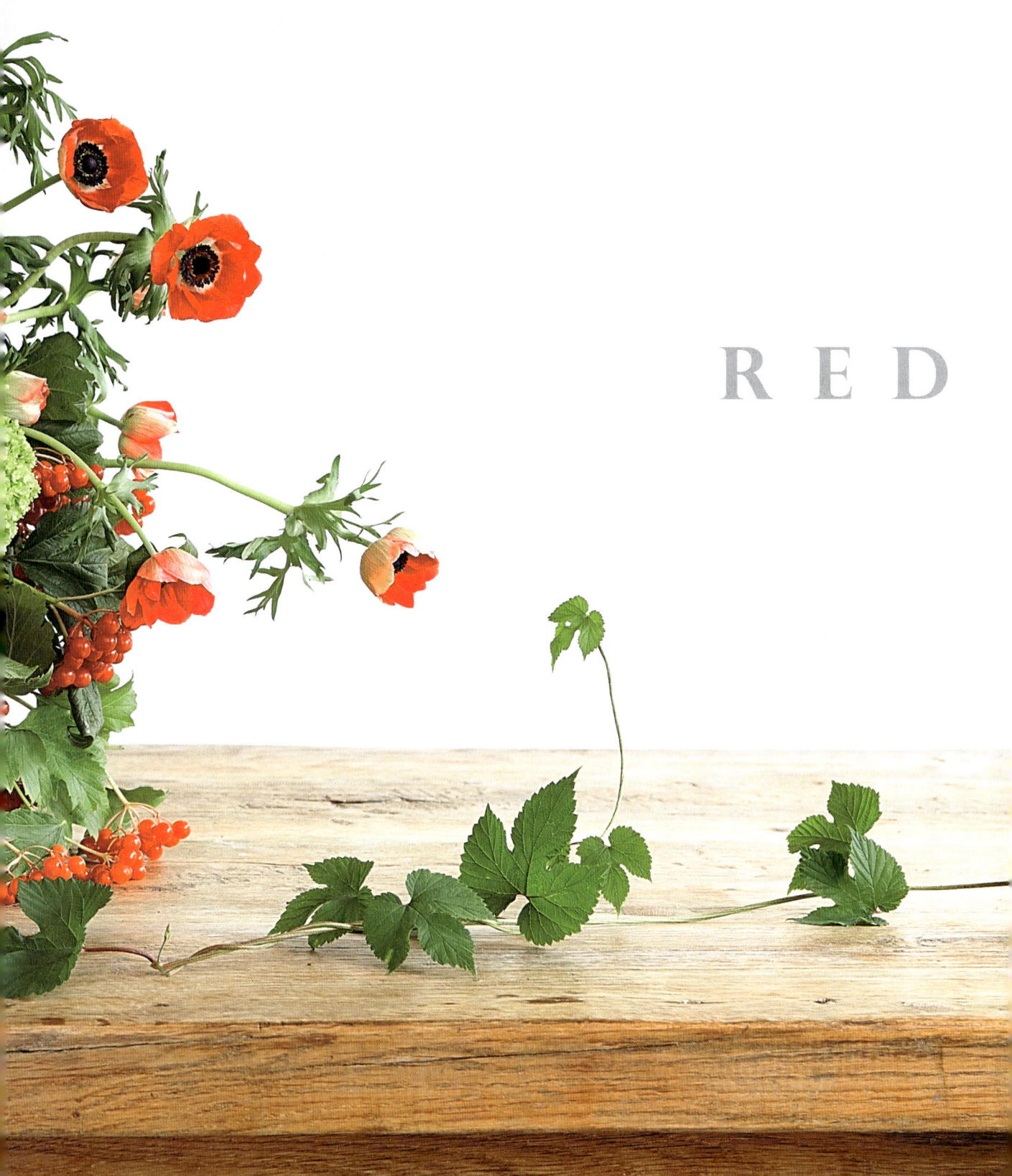

Chapter 1

RED

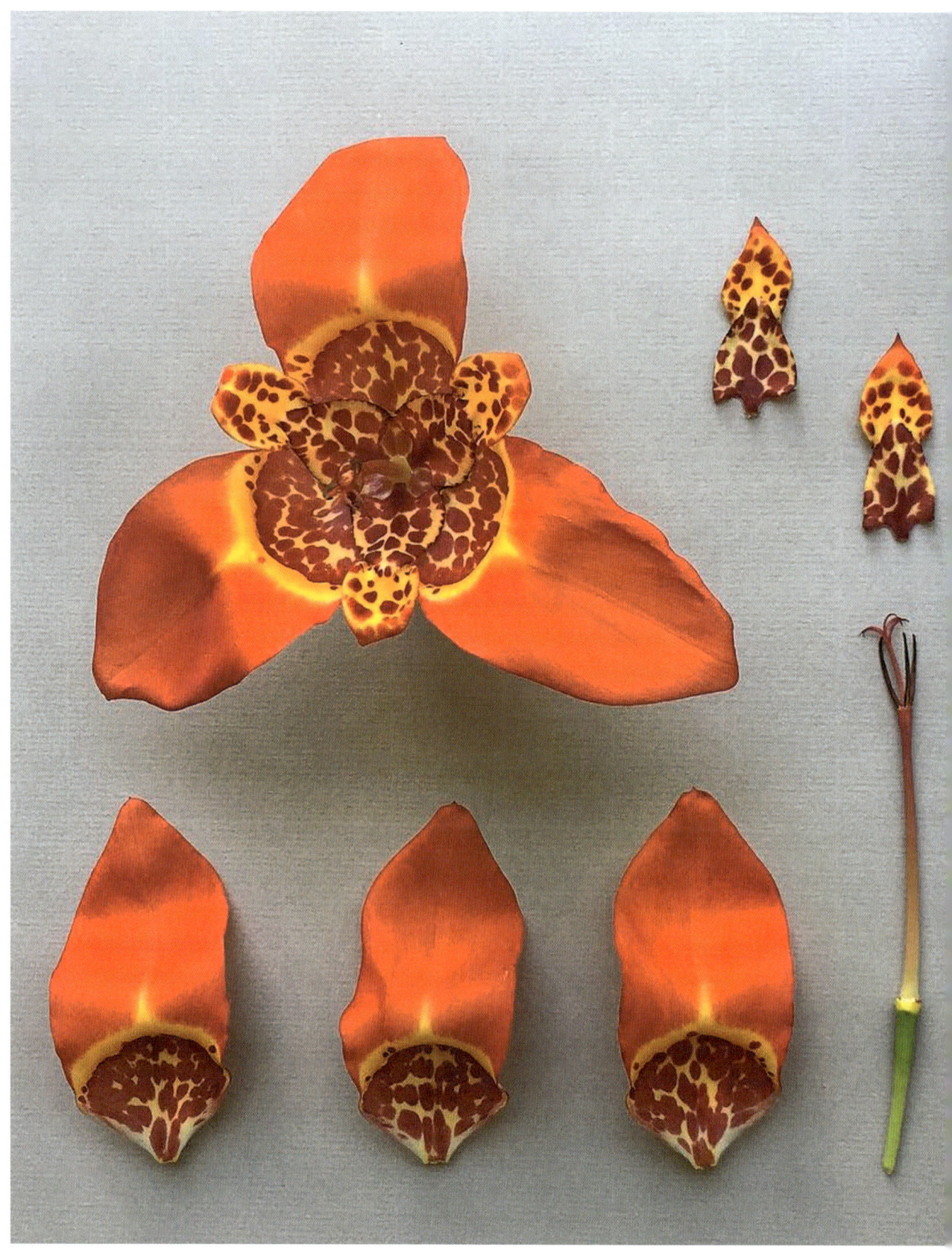

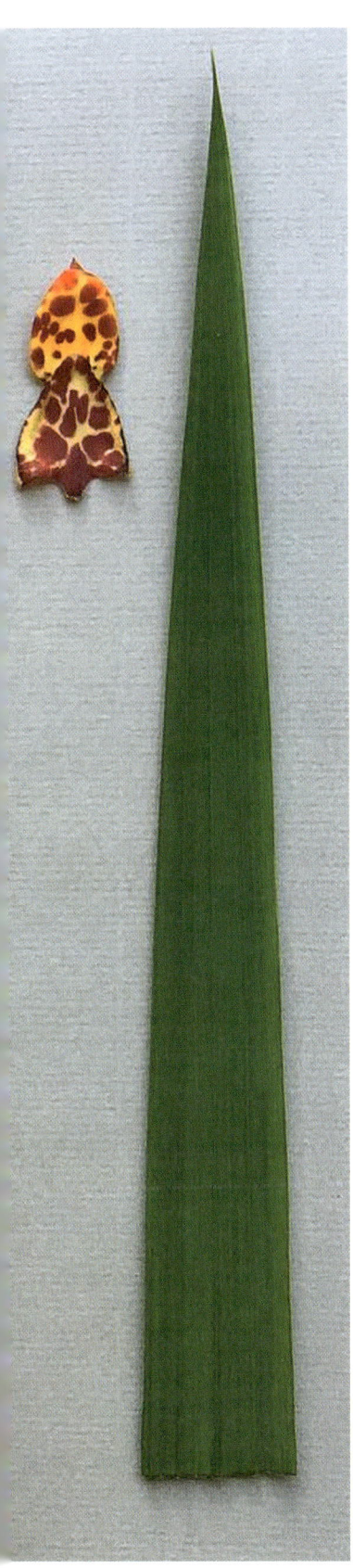

PEACOCK FLOWER

Latin name *Tigridia pavonia*
Photograph shows Flower, small petals, leaf, pistil, large petals.
Collected Teviotdale, Harare, Zimbabwe, 2016
Other common names Tiger flower, Mexican shell flower
Family *Iridaceae*
Origin Seasonally dry sands and grass lands of Mexico, Guatemala, El Salvador, Honduras.
Description A bulbous, summer-flowering perennial that grows up to 45–60cm tall.
Leaves Pleated sword-shaped leaves form in a broad, fan arrangement.
Flowers These are 7–15cm wide and come in all shades of orange, pink, yellow, white, red, with contrasting yellow markings. Flowers open early in the morning and close when dusk begins to fall.
Fruit Not ornamentally significant.
Nomenclature *Tigridia* means 'tiger like' alluding to the colour and spots of some of the species and *pavonia* means peacock like.
Uses Planted as an ornamental, the roots are also edible and were eaten by the Aztecs of Mexico.

As a child I spent many a weekend at Teviotdale farm on the outskirts of Harare with my friend Ryan Carney. His mother Sally grew up on the farm, and the Carneys lived in a cottage adjacent to the main homestead. Ryan and I rode horses in the bush, worked with clay in Sally's studio, and generally got up to mischief. Ryan's father, Danny, an author, was a terrific storyteller and would have us enraptured with tales, most often taken from his latest book. A favourite was 'Wild Geese', a story about mercenaries in Africa, later made into a film with Roger Moore and Richard Burton. I collected this *Tigridia pavonia* from the homestead garden, which is a riot of 1950's colour.

An enduring Teviotdale flower memory from my childhood is of a 'paper' chase that Sally set up for one of Ryan's birthdays. She picked *zinnia* flowers from the garden and laid a vibrant floral trail down the valley that led us to a picnic laid out on the edge of the Mazowe Springs. As a result of that trail *zinnias* still flourish in the valley, their bright blooms incongruous in the bushveldt.

CHINESE HIBISCUS

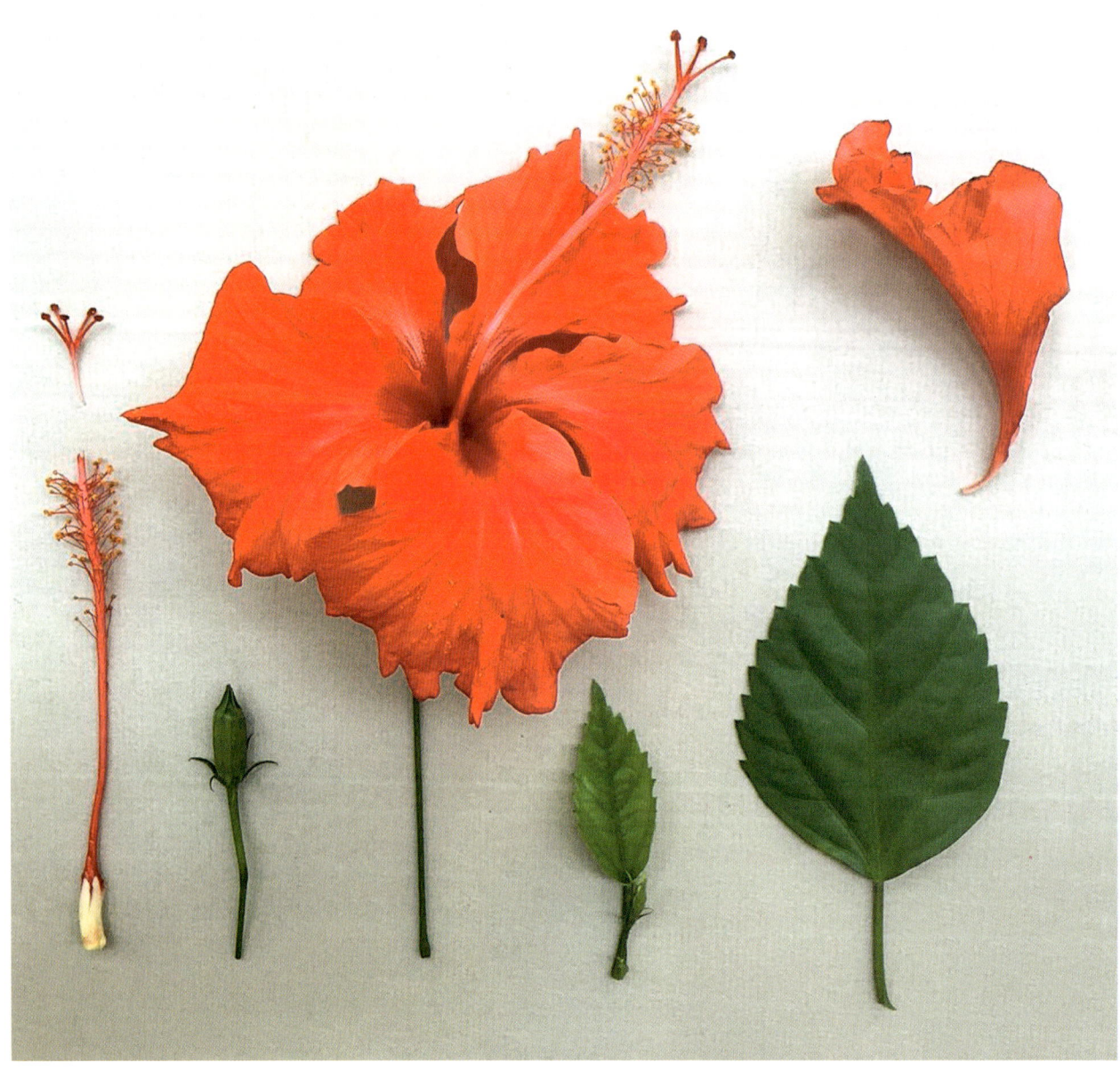

I flew to Martinique to meet up with a scientist friend, Anna, who
had given up a career in the petroleum industry to sail around
the Caribbean. Once there we spent a week in a beach villa in Le
Diamant. Armed with a couple of books on native flora and fauna we
spent a week barefoot, traversing the island. I collected this specimen
on a roadside near the house.

Latin name *Hibiscus rosa-sinensis*
Photographs show Right: stigma, buds, flower head, leaf, petal. Opposite: stigma, filament with anthers, bud, flower, petal, mature leaf, new leaf with bud.
Collected Right: Ibiza, 2106. Opposite: Le Diamant, Martinique, 2015
Common names Hawaiian hibiscus, Rose of China
Family *Malvaceae*
Origins Warm-temperate, tropical, and sub-tropical areas of East Asia.
Description An evergreen shrub or small tree that grows up to 2.5–5m tall, with a spread of 1.5–3m.
Leaves Glossy, dark-green leaves have toothed, or dentate margins, and measure 10–20cm in length.
Flowers Solitary, short-lived, funnel-shaped flowers are 10–20cm wide and have five toothed petals, fused at the base to the central column. The five-lobed central staminal column has multiple yellow anthers near its tip.
Fruit Forms ovoid, five-celled, dehiscent capsules 20–25mm long. The capsule opens when it is mature, releasing many seeds.
Nomenclature *Hibiscus* is Latin, from the Greek *hibiskos* referring to another plant in *malvaceae* family – the marsh mallow. *Rosa-sinensis* means 'rose of China'.
Uses The plant is used in ayurvedic and herbal medicines. The edible flowers are used in chutneys, soups, and curries.

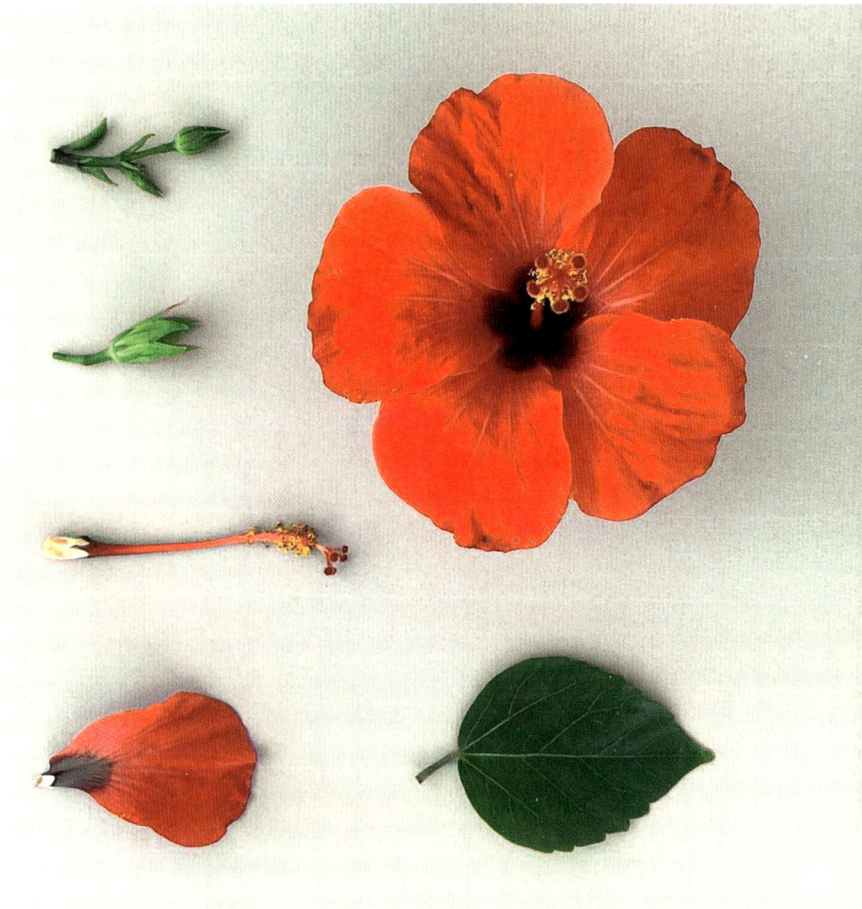

When my parents first married, they built a house on land next to my grandparent's home, and planted a red hibiscus hedge between the two. As kids, we would pick the flowers, which we named 'queens', and deconstruct them. First we removed her 'crown' (stigma), then her 'wig' (anthers) and 'skirt' (petals), and finally her 'undergarments' (calyx).

FLAMING PARROT TULIP

Latin name *Tulipa* 'Flaming Parrot'
Photograph shows Bud, leaf tip, flower, anthers, tepals.
Collected London, UK, 2015
Family *Liliaceae*
Origin Central Europe – warm, dry positions from sea levels to alpine areas.
Description Bulbous perennial that grows up to 60cm; bulbs can be up to 10cm in diameter.
Leaves Strap-shaped leaves are arranged alternately at the base of the stem.
Flowers Large, cup-shaped flowers bloom in spring and early summer. 'Flamed' yellow and red, they have purple/black anthers. Petals are fringed and open out as flower matures.
Fruit Forms globose capsules with a leathery covering, which contain flat, disc-shaped seeds in two rows per chamber.
Nomenclature It is thought that *tulip* derives from the Persian word for turban *dolband* or Turkish *tulbent*, as the flower resembles a turban in shape. 'Flaming Parrot' is the cultivar name.
Uses Decorative ornamental.

The attractive striation and irregular petal edge of this Parrot tulip are the result of breeding stable specimens of tulips with the 'Tulip-breaking' virus, which causes colour breaking. The concept of using a virus or a disease to produce more aesthetically pleasing progeny is an interesting topic. The characteristic and desirable ridge of hair running down the back of a Rhodesian Ridgeback dog, is in fact caused by a mild form of spina bifida.

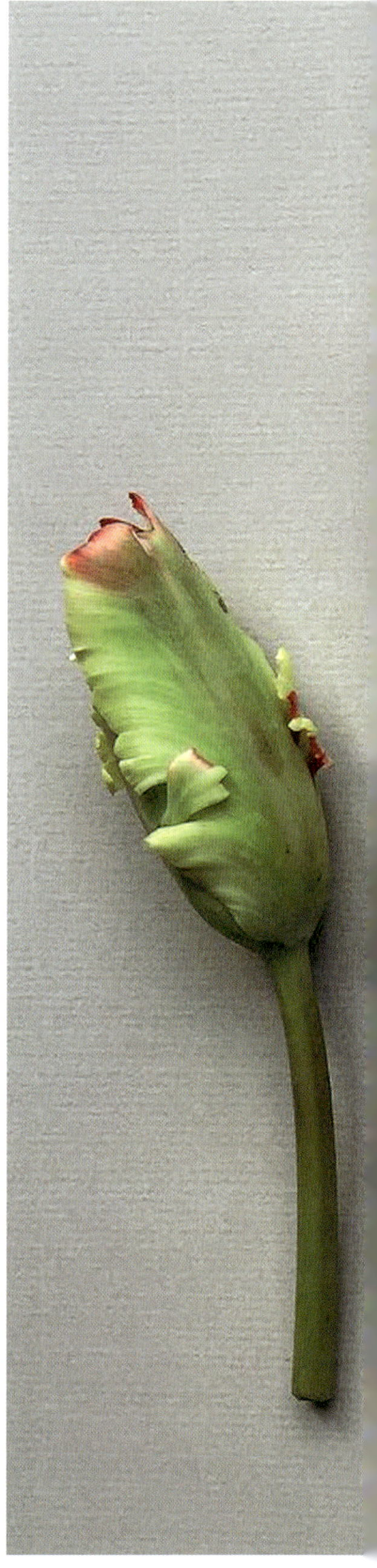

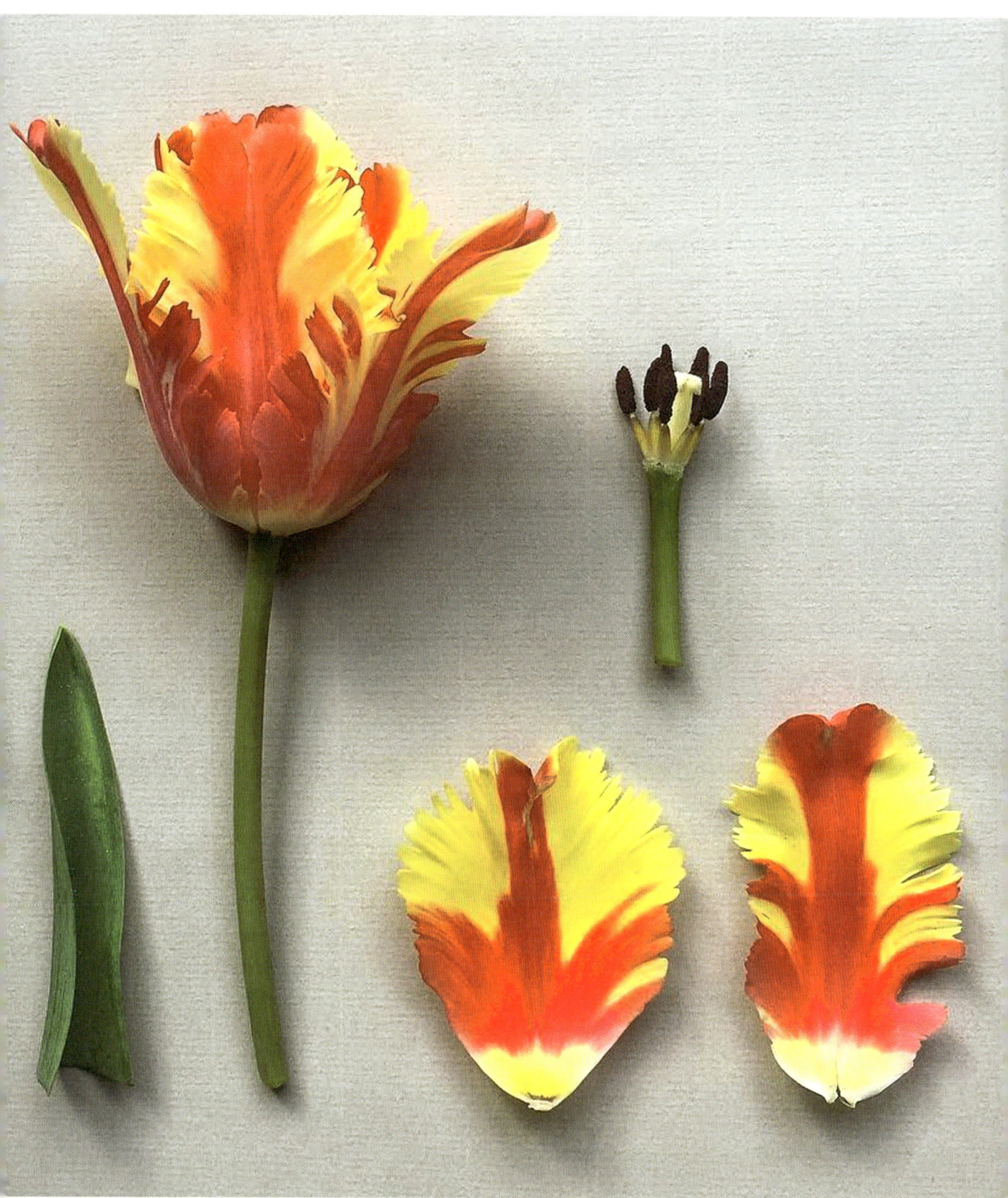

LADY'S SLIPPER VINE

Latin name *Thunbergia mysorensis*
Photograph shows Vine with leaves and inflorescences showing open flowers and buds, inflorescence stem, stamen with anthers, calyx, stigma, bud, fruit.
Collected Pointe aux Piments, Mauritius, 2017
Other common names Mysore clock vine, Indian clock vine
Family *Acanthaceae*
Origin Tropical and subtropical southern India.
Description Vigorous, evergreen summer-flowering, herbaceous climber with woody twining stems, that grows to about 6m.

Leaves Features long, ovate, lanceolate, medium-green, 12–15cm-long leaves with irregularly dentate margins.
Flowers Vine bears large (5cm long), funnel-shaped, two-lipped flowers (red-brown on the outside and golden yellow inside. They form on pendulous racemes that are 35–100cm in length.
Fruit None
Nomenclature The genus is named after Swedish physician and botanist Carl Peter Thunberg (1743–1828), a protégé of Linnaeus. The epithet *mysorensis* refers to Mysore, a city in southern India.
Uses Decorative ornamental.

Chinakwaremba, the house my parents built when they were first married, had a pergola covered in this vine. The flowers hung down on ever-elongating slender inflorescences, sometimes up to a metre long. When I was a baby these cascades of golden flowers were my first mobiles and my first botanical crush.

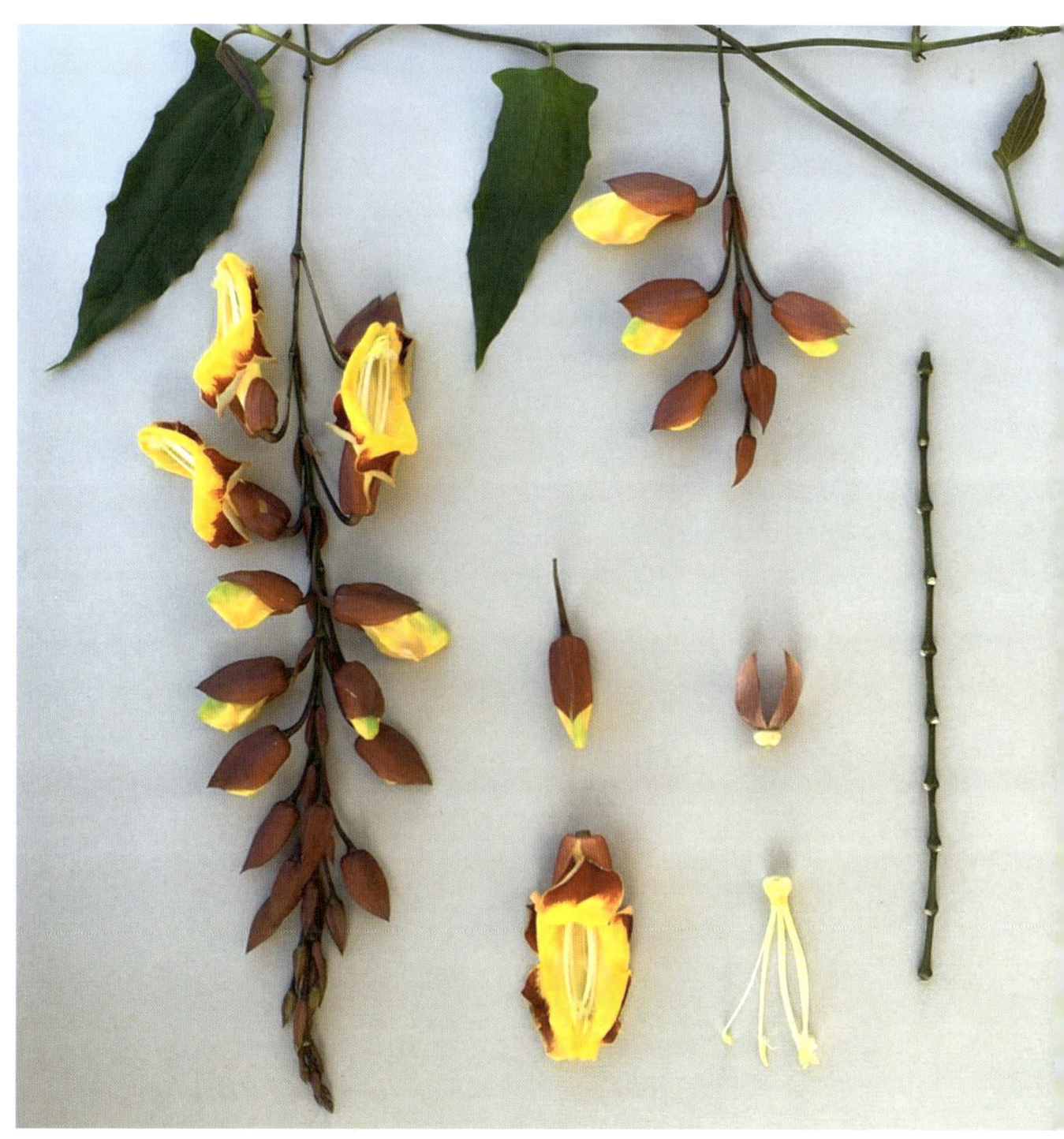

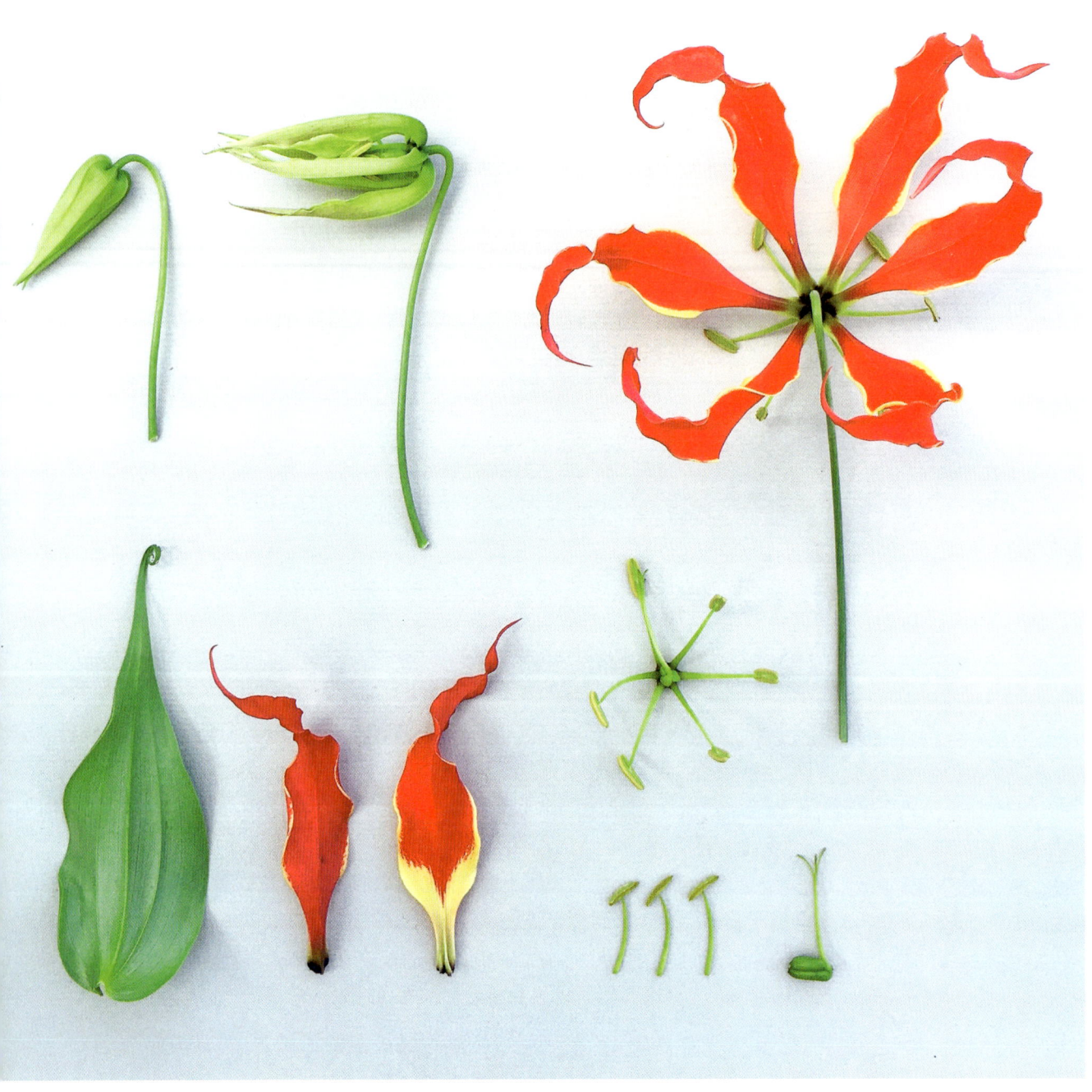

FLAME LILY

Latin name *Gloriosa superba*
Photographs show Opposite: Bud, immature flower, flower, ovary with pistil, stamens, corona, petals, leaf.
Right: Flower, petals, leaf, immature flower, stamens, ovary with pistil.
Collected Opposite: Domboshawa, Zimbabwe, 2016. Right: Chimanimani mountains, Zimbabwe, 2018
Other common names Climbing lily, Glory lily, Gloriosa lily
Family *Colchicaceae*
Origin Dry forests and thickets of tropical Africa and India.
Description A weak-stemmed, climbing, perennial that scrambles through other plants. Growing from a rhizome, can reach up to 3m in height. In areas with distinct dry seasons, the plant will die down after flowering and becomes dormant.
Leaves These can be 5–15cm long and 4–5cm wide. They have parallel veins and tips ending with spiral tendrils used for climbing.
Flowers Produces solitary, axillary, sometimes terminal flowers, 3.5–18cm long in different shades of red, yellow, orange, crimson, purple/mauve stripes, or fading purple; they are often bicolored. Pedicel is erect and recurved apically.
Fruit Capsule is 37–50mm long, 10–14mm in diameter and contains 4mm seeds
Nomenclature *Gloriosa* means 'full of glory' and *superba* is 'superb' describing the plant's striking flowers. The plant is known as *jongwe*, meaning rooster in the language of the Shona people.
Uses Grown as a decorative ornamental. Highly toxic, but used medicinally; plant extract is used in poisoned arrow tips.

The flame lily is the national flower of Zimbabwe. The UK's Queen Elizabeth II was given a diamond flame lily brooch for her 21st birthday in 1947. It was a gift from the children of what was then known as Southern Rhodesia, who each donated a 'tickey' ($2^1/_2$ cents) towards the piece. She is often seen wearing it all these years later.

After the rains, the flowers of the red flame lily embroider the newly verdant countryside. Rather vexingly I have difficulty spotting them, because I have deuteranopia, a form of colour blindness, which means that although I can distinguish red and green individually, if they are side by side and viewed from a distance, my eyes only register the green, but close up I can separate them again.

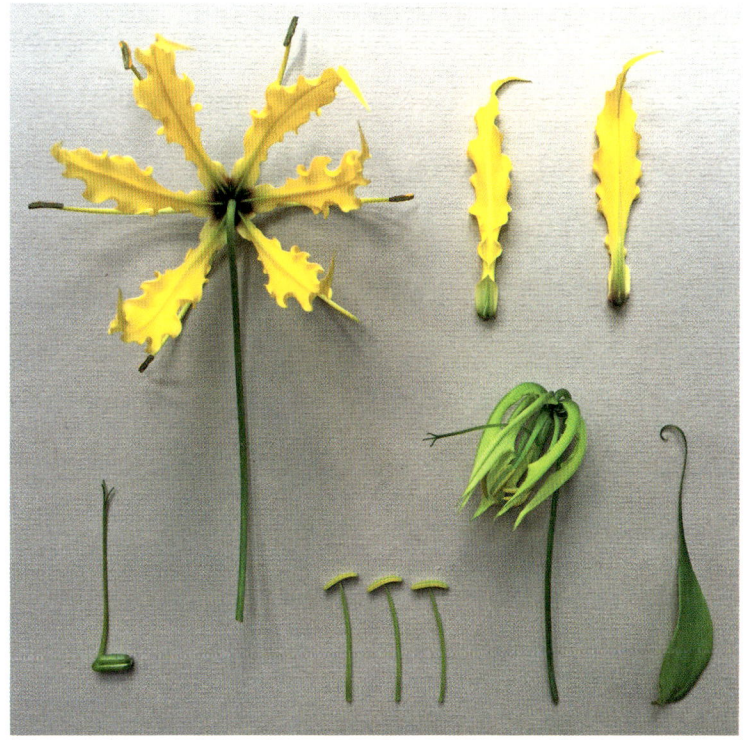

I collected this yellow variety whilst hiking in the Chimanimani mountains in the Eastern Highlands of Zimbabwe. *Gloriosa superba* exists in many colours and plant varieties. A form with wide, smooth plum-coloured petals, is found mostly in the South Africa's Lowveldt. Orange forms are found in the western Lowveldt and Botswana. These pure yellow forms are found in the Eastern Highlands. Some varieties were formerly described as separate species, but they are all now regarded as belonging to one highly variable species, characterised by the tendril-bearing leaves.

ABYSSINIAN CORAL TREE

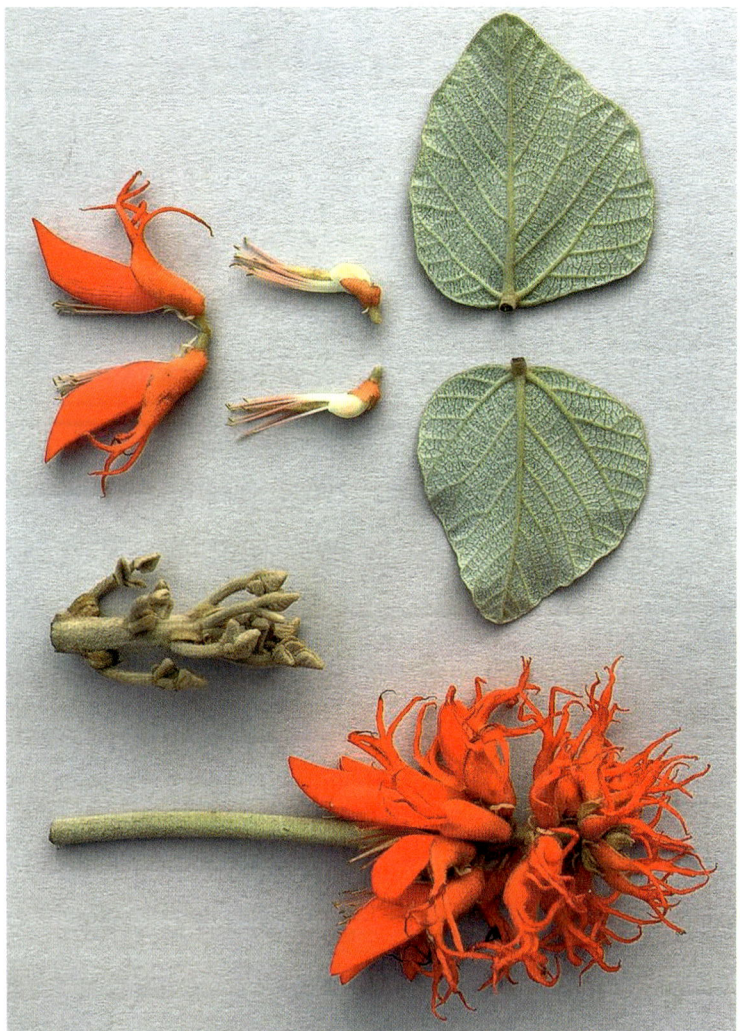

Latin name *Erythrina abyssinica*
Photograph shows Leaf shoots, flowers, stamens with filaments, underside of leaves, mature inflorescence.
Collected Maduma, Harare, Zimbabwe, 2015
Other common name Red hot poker
Family *Fabaceae*
Origin Subtropical and tropical areas of eastern Africa, eastern DRC (Democratic Republic of the Congo), and southern Africa.
Description Medium-sized, deciduous, late-winter-flowering tree growing up to 15m, and occasionally 25m tall.
Leaves Forms trifoliate leaves that are broadly elliptic; the terminal leaflet grows up to 15cm, while the lateral leaflets are about 10cm. Young leaves are hairy and may have prickles scattered on veins on the underside.
Flowers Orange- or scarlet-red flowers are borne on a pseudo-raceme up to 20cm long in the dry season before leaves develop.
Fruit Produces dark brown to black, narrow, cylindrical seed pods that can be up to 10cm long and contain red seeds with a back spot on one side.
Nomenclature *Erythrina* derives from the Greek word *erythros*, meaning 'red' describing the red flowers of some of the species. *Abyssinica* denotes the first specimen brought to Europe from Ethiopia, known then as Abyssinia.
Uses Medicinally for a variety of conditions including bilharzia, conjunctivitis, and chlamydia. Cork from the bark is used as floats for fishing nets. The seed are used as beads to make necklaces and bracelets.

In the valley below Maduma there is a stand of ancient *Erythrina abyssinica* that pre-date the house. My father built an epic tree house high up in the branches of one of them for my brother and I. It was particularly spectacular when the bare tree bloomed with an abundance of vibrant fuzzy red flowers and again, later in the year, when it was covered with black pods containing shiny red and black seeds known locally as 'lucky beans'.

SACRED CORAL TREE

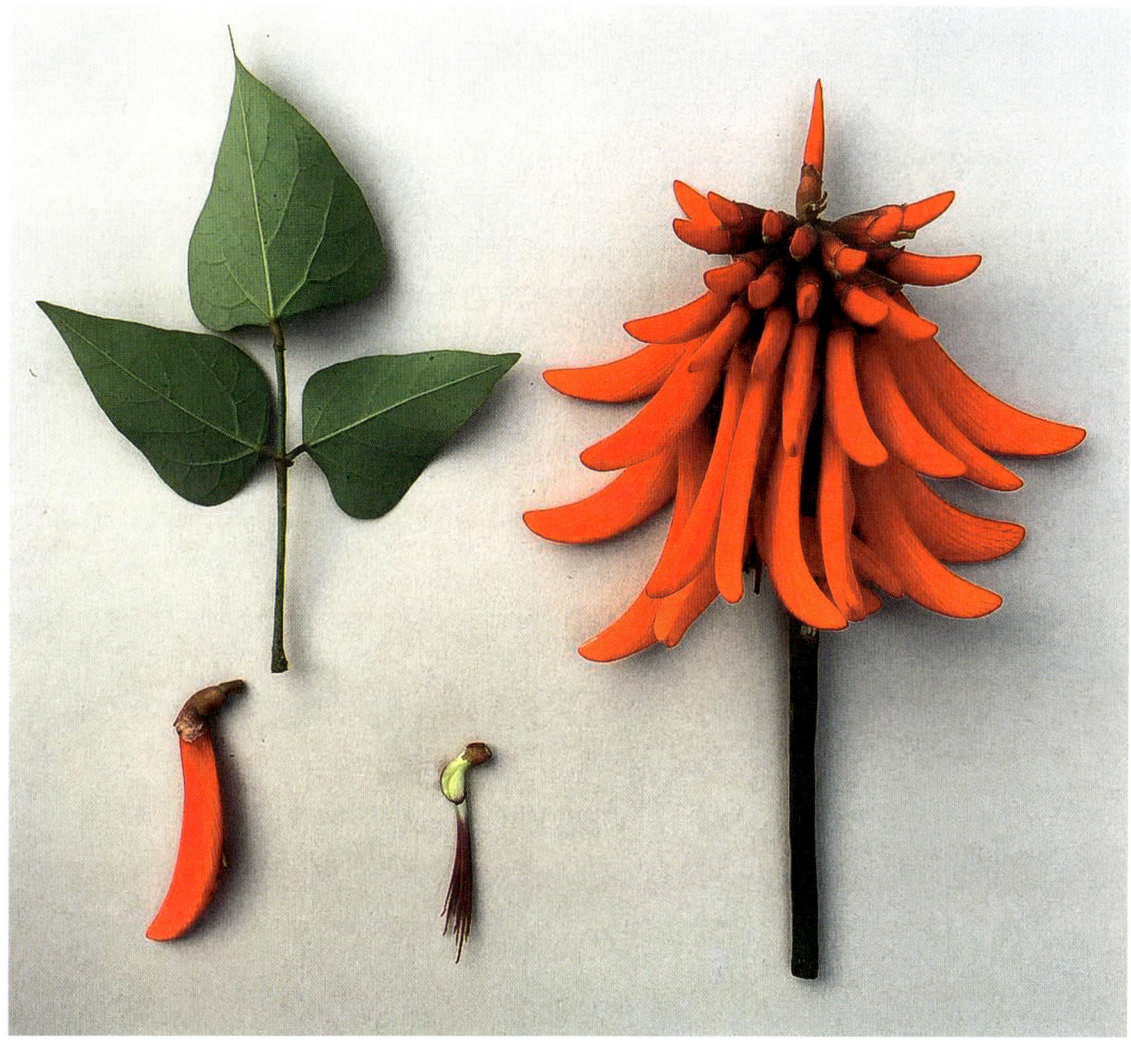

Latin name *Erythrina lysistemon*
Photograph shows Leaf, inflorescence, stamen and pistil, flower.
Collected Harare, Zimbabwe, 2015
Other common names Lucky bean tree, Transvaal kaffir boom, 'tree' in Afrikaans.
Family *Fabaceae*
Origin South Africa
Description Small- to medium-sized, deciduous tree growing up to 12m tall.
Leaves Trifoliate leaves are ovate; terminal leaf grows up to 10cm long, while the lateral leaflets are only about 7cm.
Flowers Dense heads (up to 9cm long) of bright scarlet-red flowers are borne on thick stalks in the dry season.

Fruit A black slender pod growing up to 10cm, which contains bright orange-red seeds with a black spot on one side.
Nomenclature *Erythrina* derives from the Greek word *erythros*, meaning 'red' describing the flowers of some of the species. *Lysistemon* is Greek, meaning 'with a loose or free stamen' referring to the stamen structure.
Uses Mainly as a food source for animals: black rhinos, elephants, and baboons eat the bark; bush pigs dig up the roots and eat them; vervet monkeys nibble on the flower buds. Birds, including woodpeckers, nest in the trees, and bees use the hollow trunks as hives.

Covered in scarlet flowers, these trees provide a welcome splash of colour against the barren, dry, winter landscape.

FIREWHEEL TREE

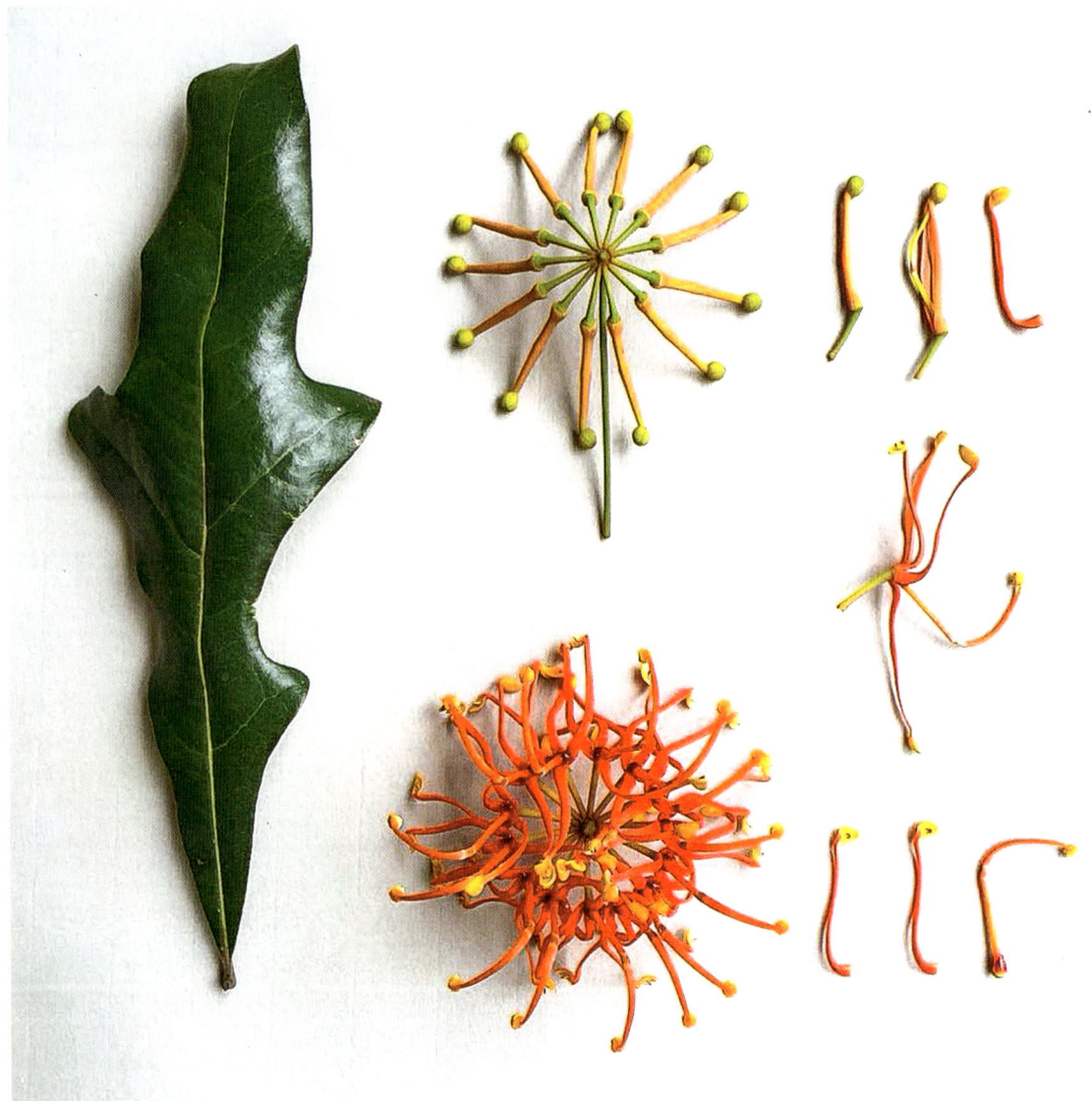

Latin name *Stenocarpus sinuatus*
Photograph shows Leaf, umbel of buds, open umbel, unopened flowers, anthers, stigma, open flower.
Collected Harare, Zimbabwe, 2016
Other common names White beefwood, Queensland firewheel tree, Tulip flower, White oak, White silky oak.
Family *Proteaceae*
Origin Rainforests of Malaysia, New Caledonia, Australia – *S. sinuatus* grows in rainforests of north-eastern New South Wales and eastern Queensland.
Description Slow-growing, medium to large, evergreen, columnar tree that grows up to 50m. It has grey bark, but young branches are brown.

Leaves Forms oblong to lanceolate leathery leaves with entire or wavy margins, that can be 15–25cm long.
Flowers Produces wheel-like umbels up to 10cm in diameter in summer. Umbels consist of 12–15 flowers that radiate from the peduncle in a single row.
Fruit Bears boat-shaped follicles with short hairs that grow up to 6cm. Each follicle contains many thin seeds up to 12mm in length.
Nomenclature *Stenocarpus* comes from the Greek *stenos* meaning 'narrow' and *karpos*, or 'fruit' to describe the narrow seed capsules. *Sinuatus*, or 'wavy' relates to undulating leaf margins.
Uses Decorative ornamental tree.

I collected this sample early one morning on Harare's Wingate Golf course. A well-planted, veritable arboretum, the diversity of trees and birdlife there is exceptional, making morning walks a delight. I go early to avoid being hit by a flying golf ball.

FLAMBOYANT TREE

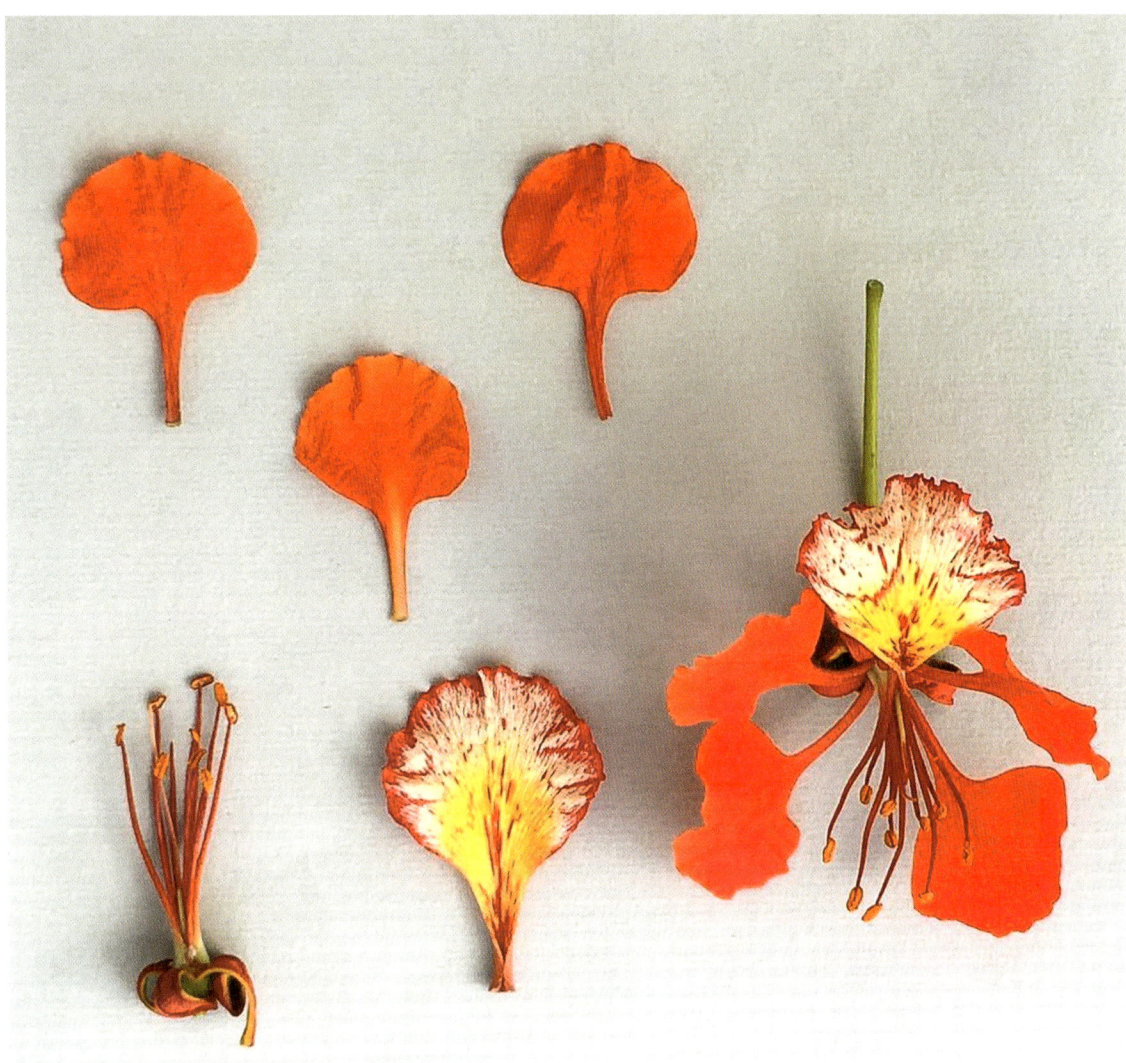

Latin name *Delonix regia*
Photograph shows Plain red petals, upright standard petal, complete flower, stamens and pistil.
Collected Mauritius, 2017
Other common names Flame tree, Royal poinciana tree
Family *Fabaceae*, subfamily *Caesalpinioiceae*
Origin Open, dry forests of Madagascar, tropical Africa, India.
Description Semi-evergreen tree (it is deciduous in areas with a long dry season) growing up to 15m with a graceful, spreading canopy.
Leaves Feathery green, fern-like leaves are up to 60cm in length, and have 20–40 pairs of primary leaflets.

Flowers In spring to summer it bears large, striking blooms, with four bright red/orange, 'clawed' petals and one yellow and white standard petal streaked red. The flowers are believed to be pollinated by sunbirds.
Fruit Large, flat, flexible, curved, green pods, which ripen brown and woody. The pods can be up to 60cm long and 5cm wide, and contain bean-like seeds.
Nomenclature *Delonix* comes from the Greek words *delos* ('conspicious') and *onyx* ('claw'), which describe its claw-like petals; *regia* is Latin, meaning 'royal' and references the tree's imposing stature.
Uses Decorative ornamental tree. The dried seed pods rattle and are used as percussion instruments in parts of Africa.

An avenue of flamboyant trees in bloom is a spectacular sight. Known as the 'City of flowering trees' there are avenues these (and jacaranda) trees all over Harare. There is a naturally occurring, yellow form, *D. regia flavia*. The only known example in Zimbabwe is in the grounds of the city's hospital, and specimens cultivated from its seed do not grow true; they always produce red flowers.

LADY FINGER BANANA

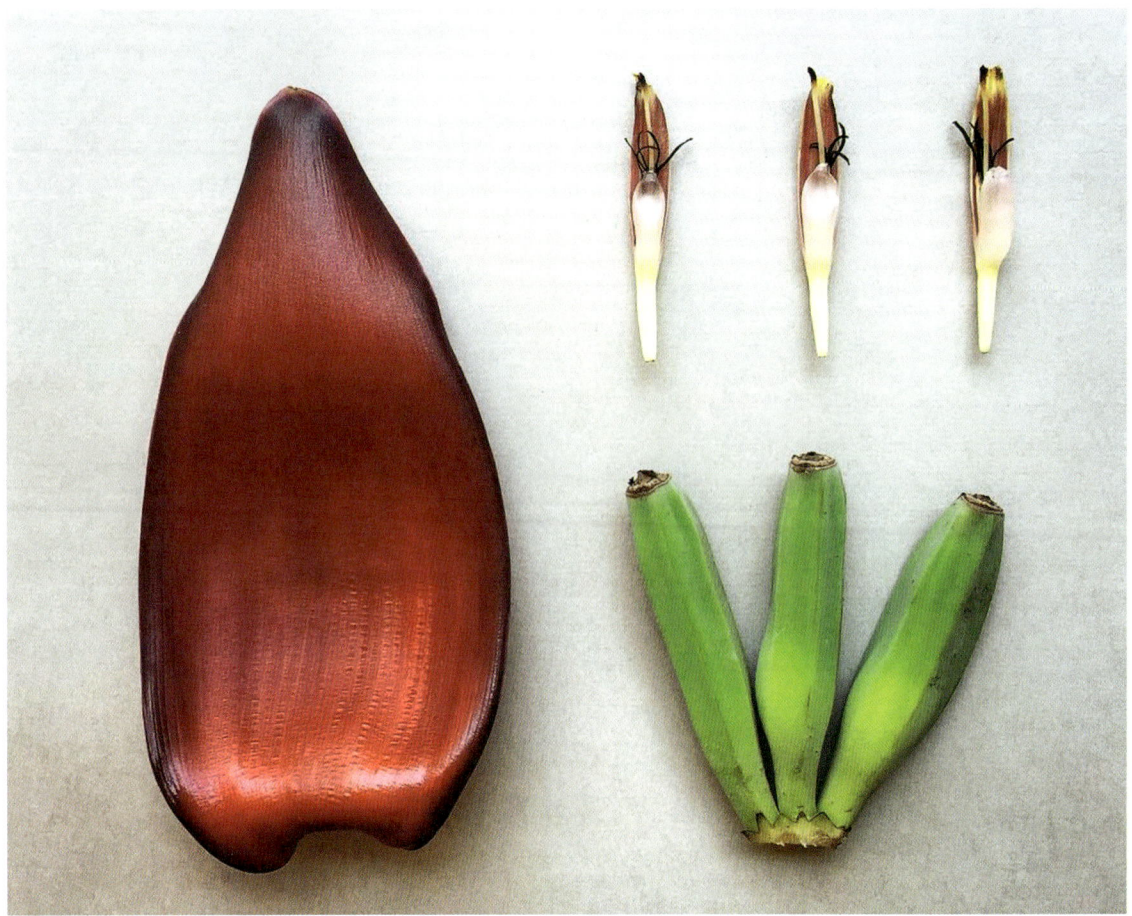

Latin name *Musa acuminata*
(syn. *M. cavendishii*)

Photograph shows Bract, male flower, berries (bananas).

Collected Maduma, Harare, Zimbabwe, 2015

Other common name Pome banana tree

Family *Musaceae*

Origin. Favours wet, tropical regions of South East Asia. Widely spread outside its native region through human intervention.

Description Evergreen, palm-like, upright-suckering perennial with false stems that can be 4–6m high and 2–3m across.

Leaves Mid-green, glaucous, oblong or lanceolate, paddle-shaped leaves have midribs and can be up to 3m in length.

Flowers In summer, a pendulous inflorescence with 2–3cm cream to yellow flowers with purple bracts (two rows per bract) emerges from the pseudo-stem. The female (pistillate) flowers are borne near the base, while the male (staminate) flowers are borne above them.

Fruit A bunch of 10–25 bananas (in clusters of up to nine) – each up to 12cm long – is borne spirally on the peduncle. The seeds of wild bananas can be up to 25 percent of the size of the mature fruit.

Nomenclature From the Latin *musa*, meaning 'banana', and *acuminatus*, means 'pointed' or 'tapering'.

Uses Cultivated in east and west Africa and across to the West Indies. Most of the bananas we see in supermarkets come from this plant. They are a staple starch in many tropical countries – the skin and inner parts of the fruit can be eaten raw or cooked. In South East Asia, banana hearts are eaten as a vegetable and said to taste like globe artichokes. The leaves are used as plates, to wrap food, and for presentation. Plant fibres are spun to make yarns for weaving; the finest are in the fabric of kimonos and kamishimos in Japan.

My grandfather loved lady-finger bananas, and devoted an acre of land to growing them. The plantation was a marvellous place to play and I built a 'fort' there and spent many a night camping in it. The local fruit bats also loved the bananas, so the gardeners would harvest the fruit while it was still green and hang it from the rafters in the workshop to ripen, away from the bats. I can still recall the sweet scent of ripening bananas, mingling with the smells of fertilizer, leather, metal, and engine oil.

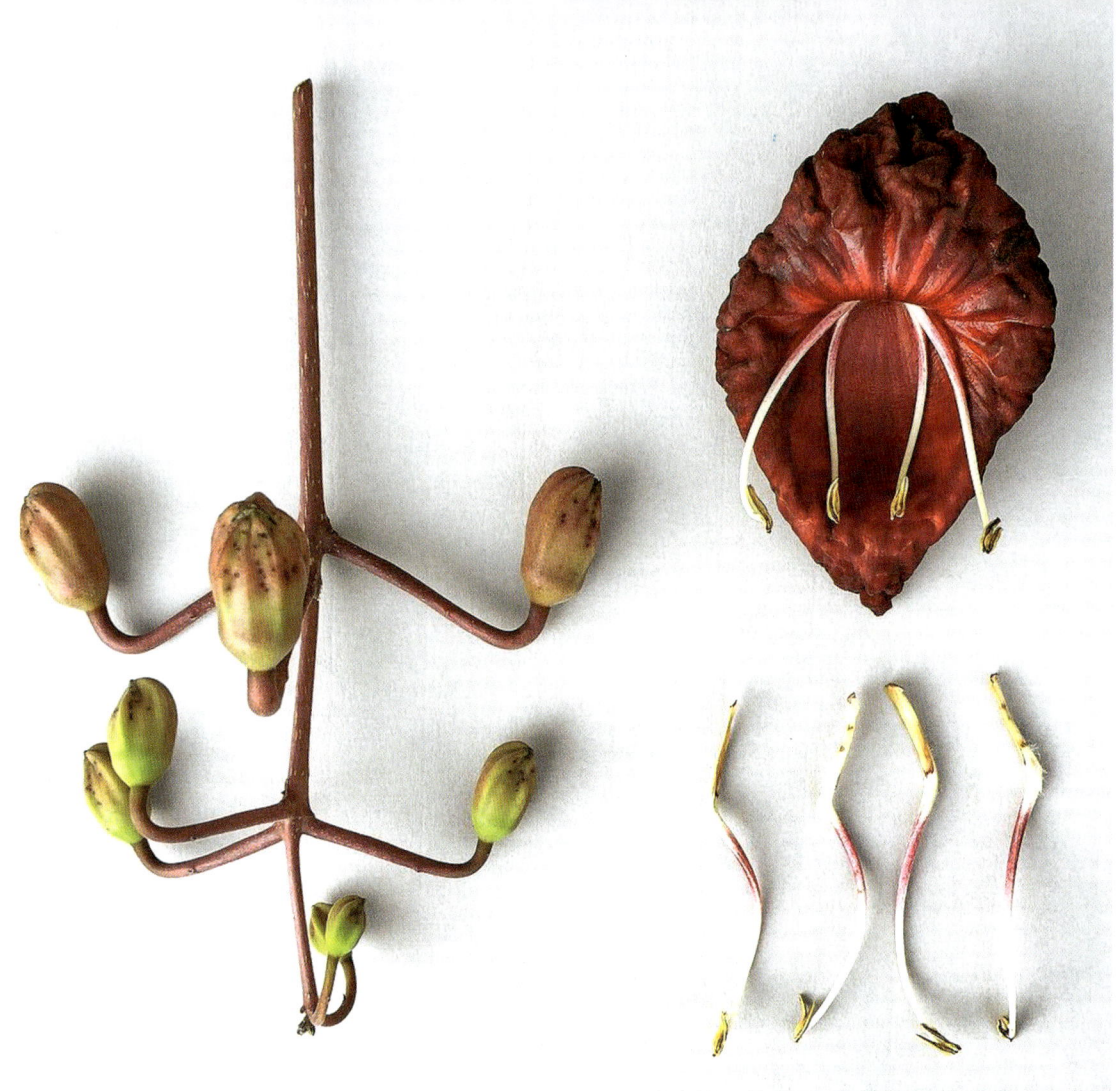

SAUSAGE TREE

Latin name *Kigelia africana*
Photograph shows Panicle of buds, open flower, stamens.
Collected Mana Pools, Zimbabwe, 2016
Family *Bignoniaceae*
Origin Woodlands and riverine fringes of tropical and north-east southern Africa.
Description Tree that grows up to 20m tall; it is evergreen in areas where rain falls throughout year and deciduous where there is a long dry season.
Leaves Compound leaves with three to five pairs of leaflets and a single terminal one. Leaves are rough and leathery, green above, and pale below.
Flowers Large, horizontal, cup-shaped flowers – 10cm in diameter – streaked green and yellow on the outside and deep maroon inside, are borne on rope-like panicles. They first open at night and emit a musky scent to attract bat

pollinators, then remain open for day-time insects. The blooms that fall in the early morning are eaten by animals.
Fruit Large pendulous, sausage-like, fruit each measuring 30–100cm and weighing up to 7kg hang from the tree.
Nomenclature *Kigelia* comes from *Kigeli-keia*, which is the Mozambican name for the tree. In Afrikaans, it is known as a *worsboom*, or 'sausage', tree, and its Arabic name (my personal favourite) translates as 'The father of all kit bags'.
Uses Fruit is eaten by animals, including porcupines, bush pigs, baboons, and elephants. The leaves and bark are used in traditional medicines. Tonga women in the Zambezi Valley apply a *Kigelia* preparation to their faces as a beauty treatment. Extract of the fruit is applied topically in treatment of psoriasis, eczema, and skin cancer.

In his diary, David Livingstone mentions camping beneath 'a giant sausage tree' shortly before he first saw the Victoria Falls. My earliest encounter with these trees and their strange fruit was as a young boy camping on the banks of the Zambezi, when we quickly learned not to sleep under them as their substantial fruit dropped unexpectedly. I was fascinated by the rubbery flowers and their odd scent. The carpets of fallen blooms were always gone by 9am having been hoovered up by passing animals.

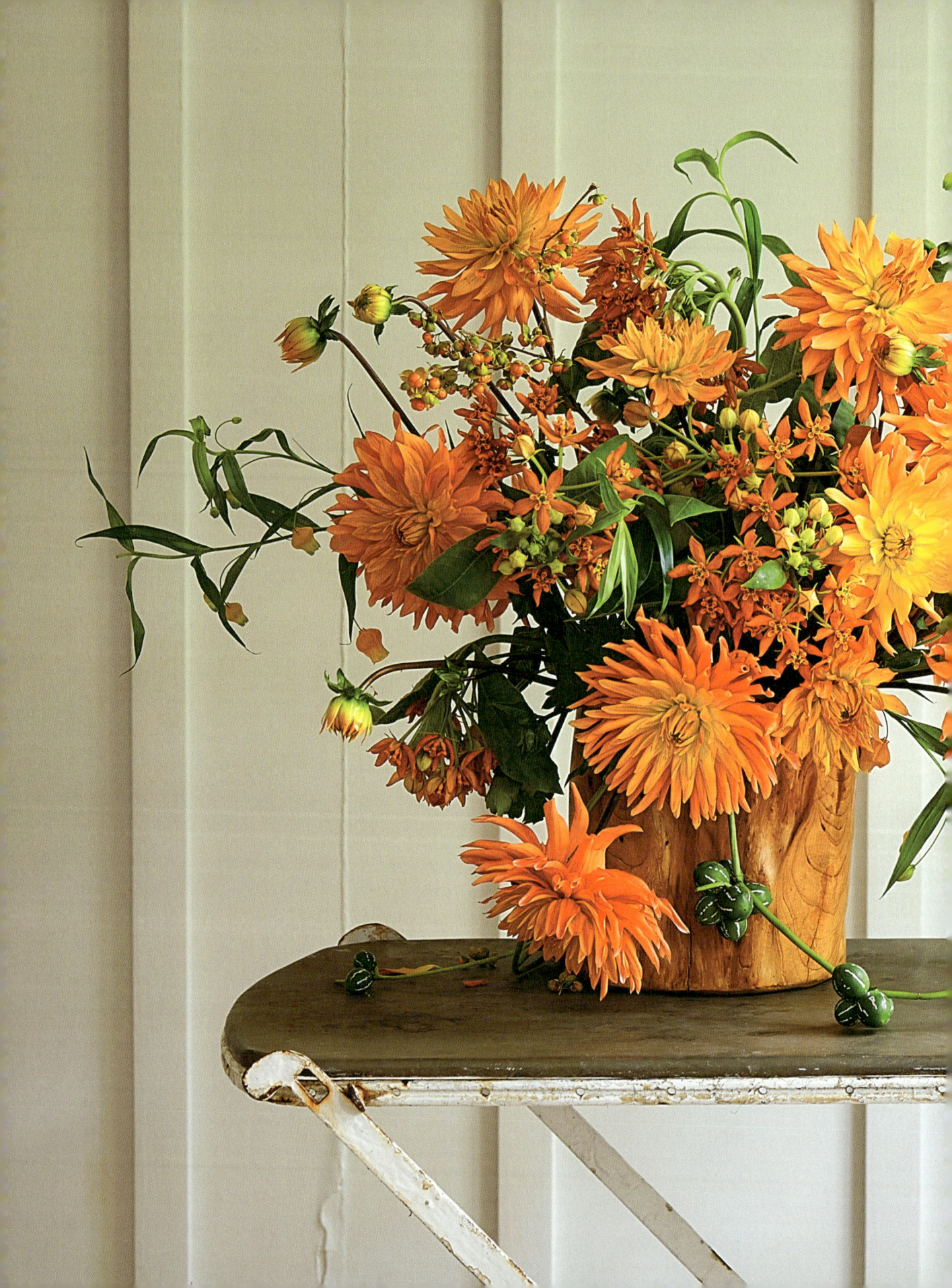

Chapter 2

ORANGE

HARLEQUIN FLOWER

Latin name *Sparaxis pillansii*
Photograph shows Leaf, petals, stamens, flower.
Collected Nieuwouldtville, Northern Cape Province, South Africa, 2015
Family *Iridaceae*
Origin South-western Africa in shale and heavy doleritic clay in waterlogged depressions.
Description An upright, cormous, spring-flowering perennial growing 15–60cm tall – the corms are usually 10–15mm across. Stems and leaves are short-lived.
Leaves Every plant has eight to ten elongated or sword-shaped, hairless, pale-green leaves, each 10–30cm long and 4–13mm wide, clustered at the base of the plant or alternately arranged along the stems. They are sheathed at the base with entire margins, pointed tips, and prominent mid-veins.
Flowers A plant bears two to seven bi- or tri- coloured funnel-shaped flowers on a zig-zag stem in a spike. Membranous bracts under the flower are 23–28mm long, have brown streaks, and are usually shallowly torn. The outer floral bracts are 20–25mm long and usually have three short cusps; smaller inner floral bracts have two.
Fruit Produces 10mm-long capsules that contain smooth, globular seeds, 2mm in diameter.
Nomenclature *Sparaxis* is from the Greek *sparasso* meaning to 'rend', or 'tear', and describes the flower's sheaths. *Pillans* commemorates the South African botanist Nevill Pillans (1884–1964) of the same name. *Tricolour* refers to the three colours of the flower.
Uses Decorative ornamental.

To celebrate my 40th birthday, I went on a two-week road trip with my parents and a friend to see the spring flowers in the Cape provinces. The blooming season begins in August and runs through to the end of September and the 30,000 hectares of the veldt are carpeted with jewel-like colours. It is weather dependent as most of the flowers are heliotropic and only open when the sun is shining, and shyly close when it is overcast. The sun shone much of the time and the spectacular flowers stretched as far as the eye could see. On overcast days, we headed to places that weren't flower-dependent. At Nieuwouldtville we saw the extraordinary and prehistoric looking forest of *Aloe dichotoma*, or Quiver trees, (the San people utilized this plant's hollowed-out branches as quivers for their arrows hence the common name). In Lambert's Bay we ate beachside, at the outdoor restaurant, Muissboskerm, feasting on freshly caught crayfish cooked over an open fire. Sublime!

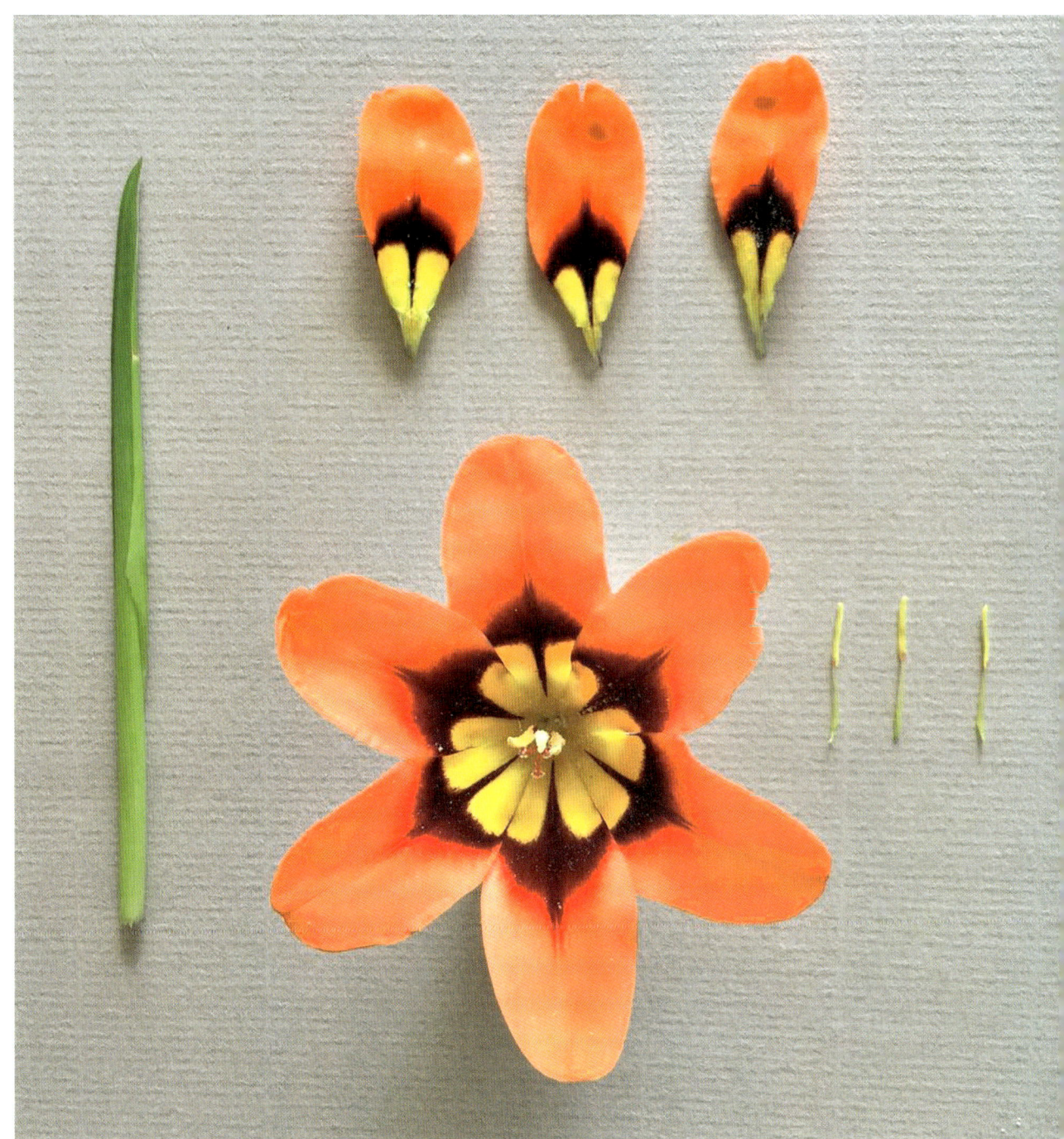

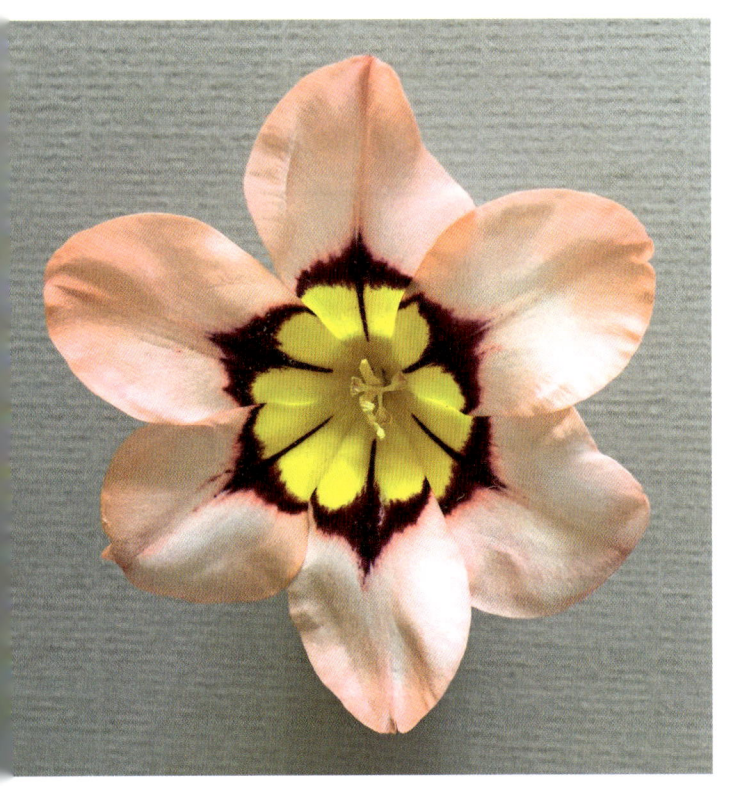

Continuing on our trip in the Cape Provinces, we headed up into the Bokkeveldt mountains, leaving the arid Knersveld valley floor behind us. We spent the night in a remote bothy atop the mountain plateau, surrounded by an abundance of wildflowers and incredible views. The following morning we visited the Nieuwoudtville Falls, where the Doring river plunges spectacularly over a cliff edge into a deep pool some 100m below. I collected all of the jewel-like Harlequin flowers on this and the previous pages from the banks of the Doring river. The photograph left shows a flower, and opposite, a leaf, petals, stamens, and a flower.

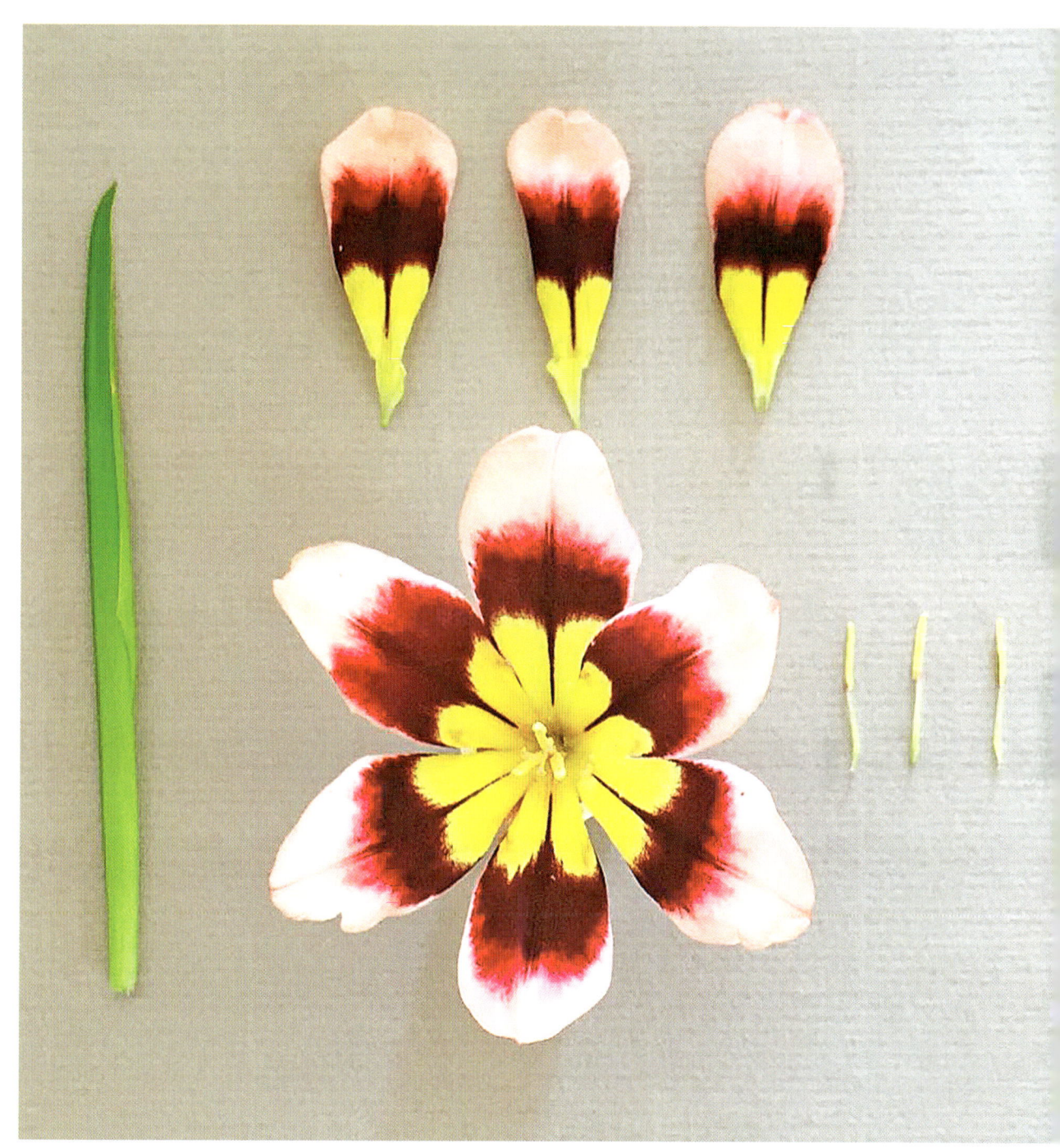

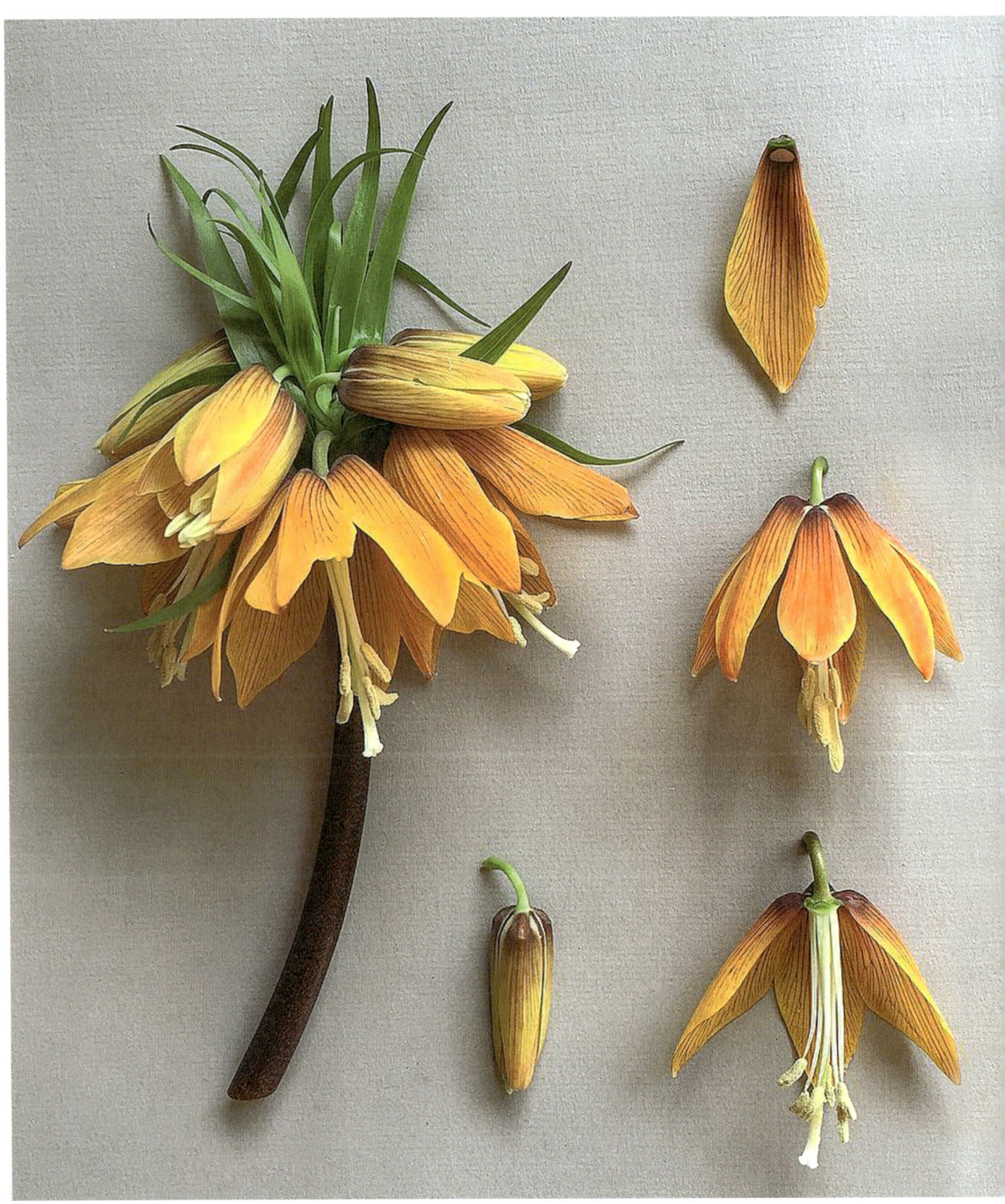

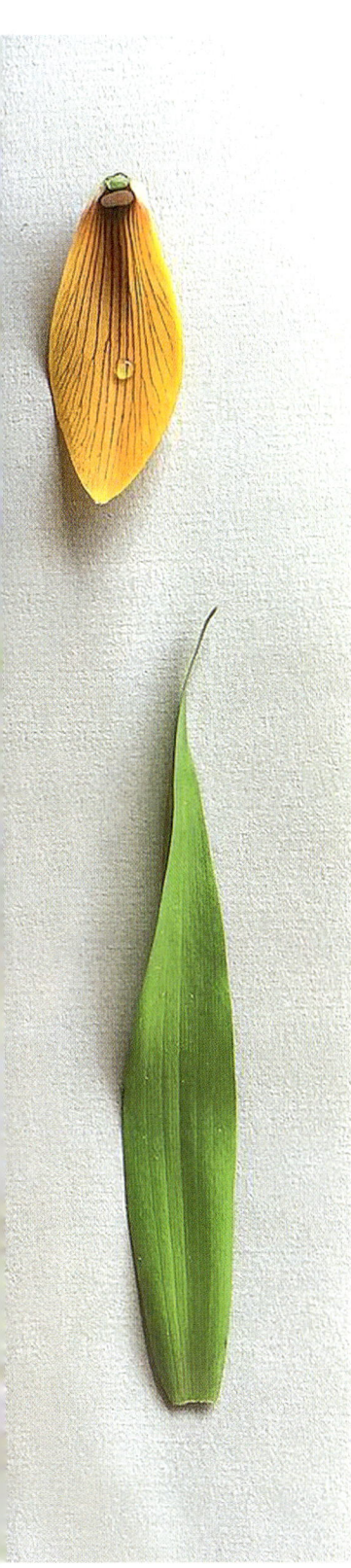

CROWN IMPERIAL

Latin name *Fritillaria imperialis*
Photograph shows Umbel of flowers, petals, leaf, section of flower showing anthers and filaments, flower, bud.
Collected London, UK, 2014
Other common names Imperial fritillary, Kaiser's crown
Family *Liliaceae*
Origin Mountainous regions of Turkey, northern Iraq, western Iran, Pakistan, Afghanistan.
Description A bulbous, herbaceous perennial that grows to heights of 40–100cm. The bulbs have a distinctive, strong smell and produce a sturdy stem that has leaves only on the lower half and a stalk topped with lily-like, bell-shaped flowers, and a tuft of leaves.
Leaves The green, simple leaves are whorled. They are lanceolate and petiolate with entire margins and parallel veins.
Flowers Umbels of five to eight pendant, usually orange, bell-like (campanulate) flowers each 4–5cm long, are arranged below the terminal tuft of leaves. Each

one has six stamina, and a tripartite stigma. They have a musky smell, are hermaphroditic, predominately self-sterile, and pollinated by bees.
Fruit Produces upright loculicidal capsules with multiple compartments.
Nomenclature *Fritillary* comes from the Latin *fritillus*, meaning 'dice-cup', possibly referring to the shape of the flowers or the dice-like, chequered markings on some the petals of some species. *Imperialis*, means 'of the emperor' referring either to the plant's stately bearing or the fact that the crown of pointed leaves atop the flowers resembles an emperor's crown, hence its common name 'Crown imperial'.
Uses The strong, vulpine odour emitted by the plants deters rodents. The bulb is said to have properties as a diuretic, an expectorant, and encourages increased lactation. Honey from the flowers is said to be emetic. However, fresh plant matter contains a toxic alkaloid called imperialine, which can cause spasms, vomiting, and cardiac arrest.

This plant exudes a strong animal-like odour, very reminiscent of fox. When I remark on this people often ask me how I know what a fox smells like? The answer is I have an observant nose and a keen sense of smell.

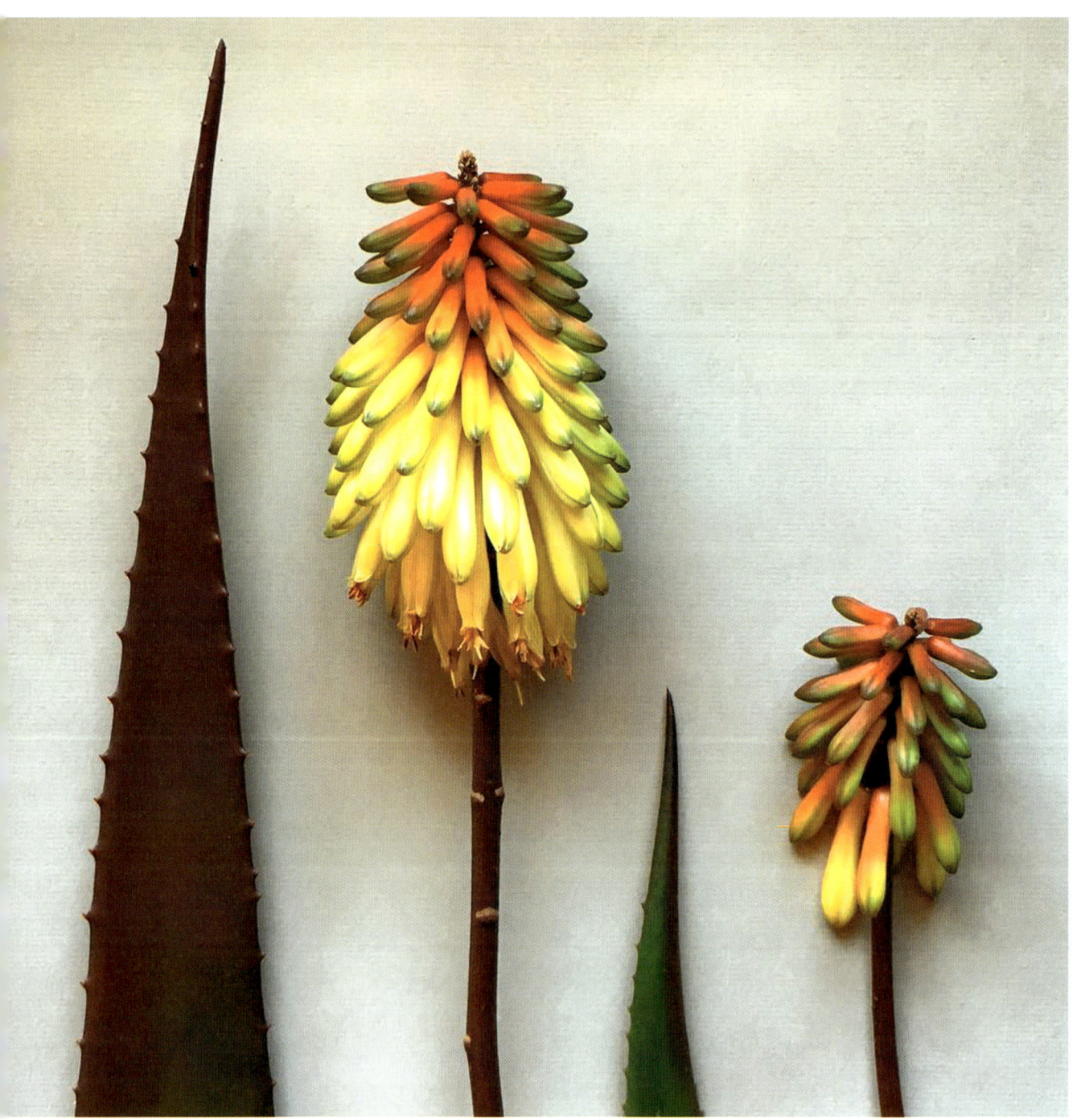

ALOE

Latin name *Aloe cameronnii* var. *bondana Reynolds*
Photograph shows Leaf, inflorescences with mature and immature flowers, small leaf.
Collected Connemara, Nyanga, Zimbabwe, 2015
Other common name Cameron's aloe
Family *Asphodelaceae*
Origin Endemic to locality of Nyanga, Zimbabwe.
Description A medium-sized, suckering aloe with many upright stems of open rosettes that grows up to 60 cm.
Leaves Thick, succulent leaves that, depending on species, can grow to 50cm in length. Medium- to dark-green, lax, narrow leaves can grow to 50cm. Arranged in a dense, rosette formation, they turn a beautiful coppery red in the dry season.
Flowers Yellow to orange and dark-red flowers are borne on a dense raceme on a stalk, or spike, that can be up to 90cm long. In *A.cameronii* the anthers and style protrude only 5mm, but var. *bondana* is distinguished by more elongated inflorescences and more clavate perianths, and the anthers and style protrude 10mm.
Fruit Forms capsules that dry out and split lengthwise to release small, black seeds; some of the seeds have translucent wings.
Nomenclature Genus *Aloe* comes from the old English *alewe*, or *alwe*, which is the name for the fragrant resin or heartwood of certain oriental trees, from the Greek word *aloē*. The variety *cameronii* is named after Kenneth J Cameron, a Scottish planter in Malawi, and *bondana* is after the Bonda Mission in Eastern Zimbabwe.
Uses The aloe species is frequently cultivated for food and medicine and as a decorative ornamental.

I collected the flowers and leaves of this spectacular yellow/bronze native aloe on the shores of Little Connemara in the Eastern Highlands of Zimbabwe. Connemara is comprised of three man-made lochs, or dams, that were the innovation of a Major MacIlwaine. Affectionately known as 'Major Mac' he was a keen angler and built the lochs to stock them with trout for sport and his portrait hangs above a fireplace at the nearby Troutbeck Hotel. Legend has it that the fire below the painting has not been allowed to go out since the hotel first opened in 1951.

AFRICAN TULIP TREE

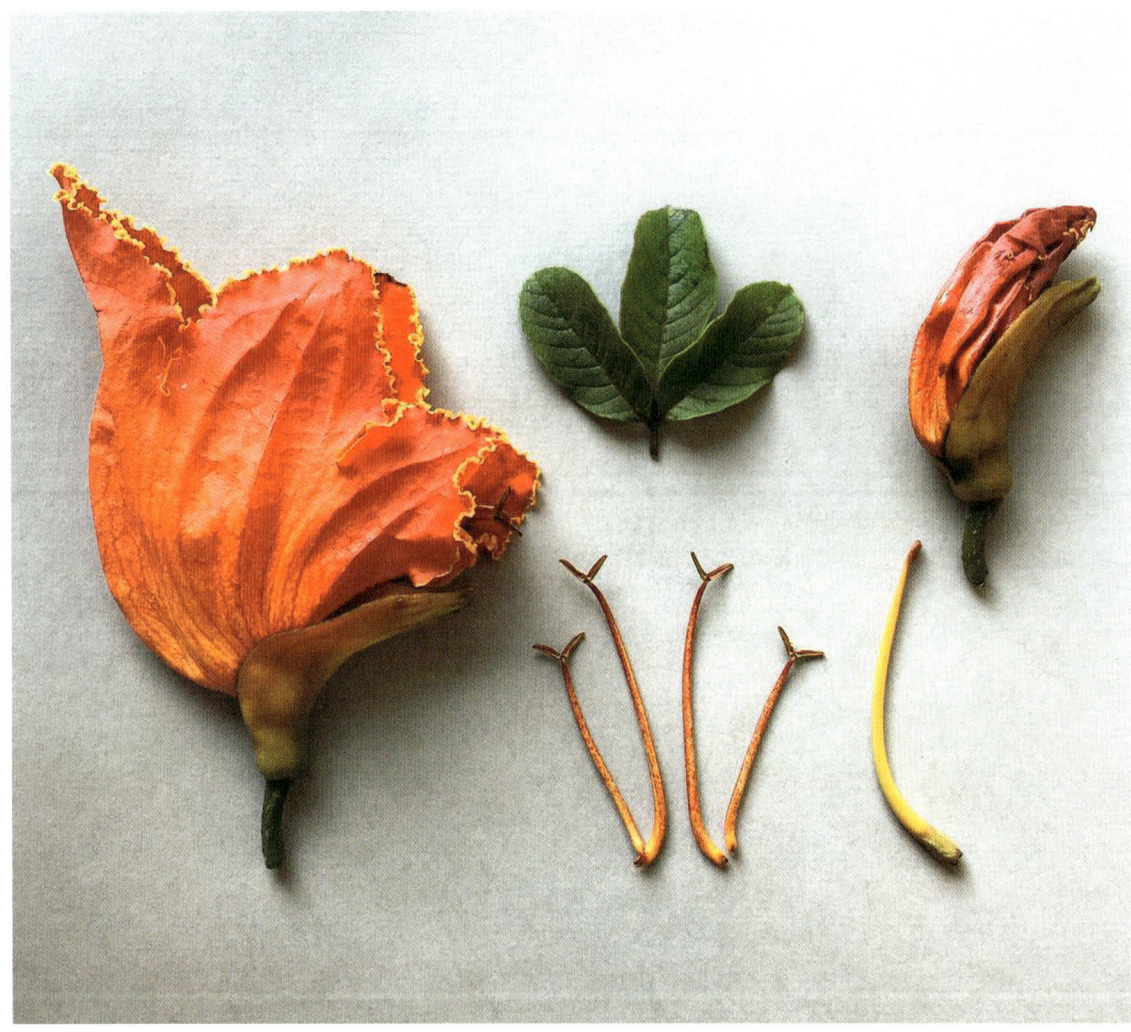

Latin name *Spathodea campanulata*
Photograph shows Flower, leaflets, bud, anthers, stamen.
Collected Harare, Zimbabwe, 2016
Other common names Flame tree, Fountain tree
Family *Bignoniaceae*
Origin Dry forests of tropical Africa – from Senegal to Ethiopia, south to Angola, southern DRC (Democratic Republic of the Congo), Tanzania.
Description Evergreen or semi-deciduous tree that grows up to 35m tall. The trunk and branches tend to become hollow as the tree ages, and branches then drop off.
Leaves Large, compound leaves (up to 50cm long) have seven to 17 leaflets – each up to 15cm long and 7.5cm wide. Leaflets are broadly elliptic or ovate, have entire margins, a sparse covering of

soft hairs, and are oppositely arranged along the stems and borne on petioles up to 6cm long.
Flowers In spring and autumn, the tree bears large, showy, bell-shaped flowers (8–10cm long) are arranged in dense clusters on stalks up to 10cm long at the tips of the branches. Individual flowers have five scarlet or orange-red lobes with yellow margins, throat, and stamens.
Fruit Produces large, elongated, flattened, pod-like capsules (17–30cm long and 3.5–5cm wide) that turn from green to brown as they mature. When ripe, they split open, releasing about 500 light, 'papery' seeds, each with a translucent membranous wing.
Nomenclature: *Spathodea* is from Greek words *spathe* ('bract') and *oida* ('to know'), referring to the calices

that around the buds. *Campanulata*, meaning 'bell-shaped', describes the flowers.
Uses Said to have medicinal properties, and is thought to be an effective malarial prophylactic. The wood is used for ornamental carvings in west Africa and for firewood and to produce charcoal in Ethiopia. It is also used to make a form of plywood distributed as African Tulip.

As many children raised in Africa know, the unopened flower buds of the African tulip tree, are filled with a pressurised, watery nectar. They make excellent water pistols when squeezed.

SNAKE TREE

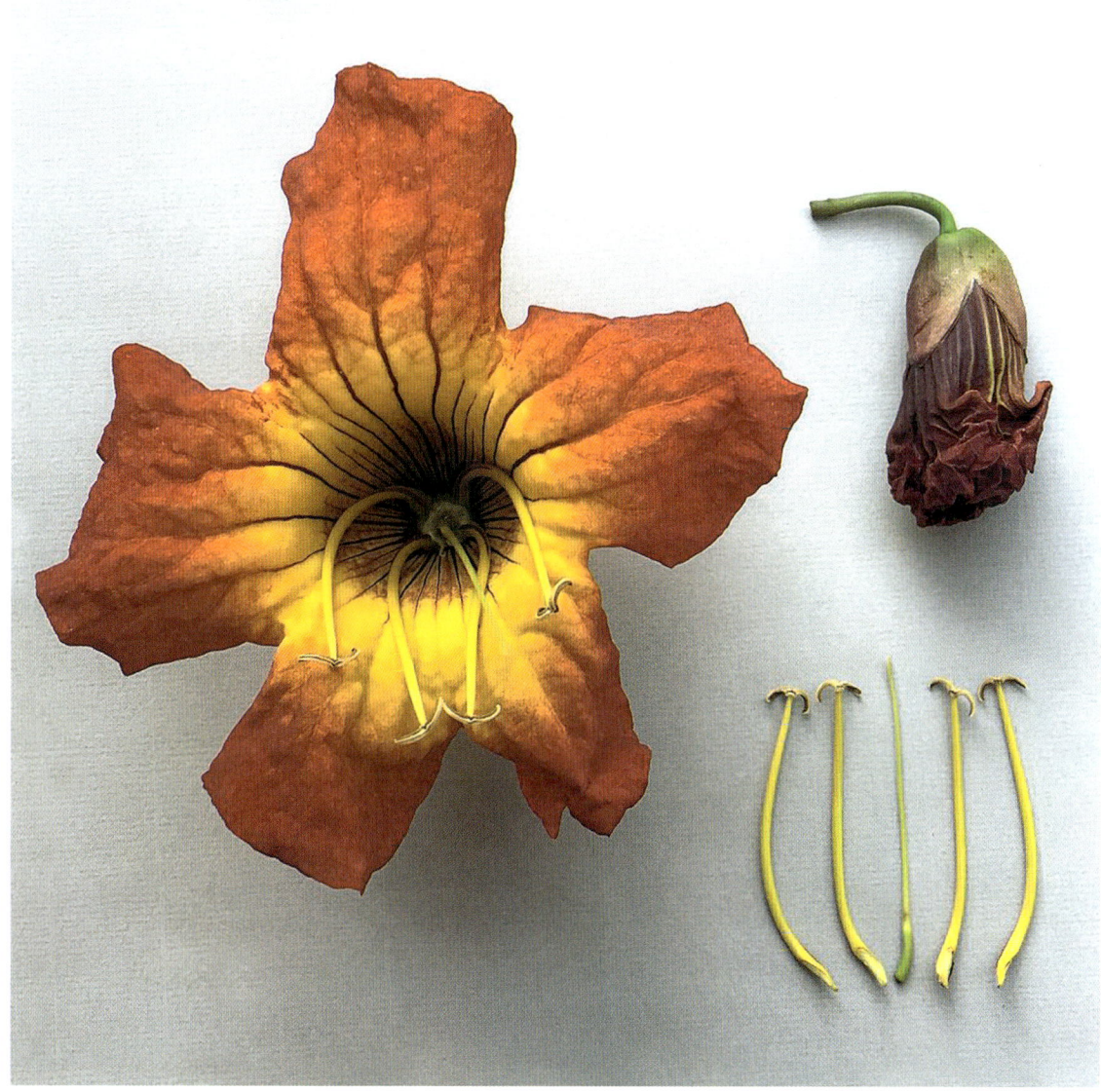

Latin name *Fernandoa magnifica*
Photograph shows Flower, bud, stamens either side of a central pistil.
Collected Borrowdale, Harare, Zimbabwe, 2016
Other common name African flame tree
Family *Bignoniaceae*
Origin Well-drained sandy soils and dry forests and woodland of East Africa – Zimbabwe to Mozambique.
Description Small, often multi-stemmed deciduous tree that grows up to 3–6m, and has grey to pale-brown bark.
Leaves Dark-green, pinnate leaves are up to 25cm long; they have four to seven pairs of leaflets and a single terminal one.

Flowers The spectacular orange, bell-shaped flowers with yellow centres are up to 10cm across and borne in abundance on racemes in spring when the trees have shed their leaves.
Fruit Produces long, slender, pale-brown, spirally twisted, or snake like, capsules 30–55cm long, containing seven to 12 winged seeds.
Nomenclature The species name '*magnifica*' meaning 'magnificent' refers to the tree's spectacular flowers.
Uses Decorative ornamental. The roots are said to have medicinal use; the wood is used to make tools and as fuel.

Rare in the wild because of the loss of natural habitat, the flowers of a snake tree are a magnificent sight. I first encountered a *Fernandoa magnifica* in bloom growing as an ornamental tree in Harare.

41

MSASA TREE

Latin name *Brachystegia spiciformis*
Photograph shows Young leaves, pods, seeds, open pods.
Collected Nyanga, Zimbabwe, 2016
Common name Zebrawood tree
Family *Fabaceae*
Origin Southern and eastern Africa – Zimbabwe, Zambia, Malawi, Tanzania, Mozambique.
Description Large shrub or small tree with a flat crown that can be 8–25m tall. Usually evergreen, it becomes deciduous in dry years.
Leaves From 5–20cm in length, leaves are made up of two to five pairs of large glossy leaflets. They are a distinctive pink, rust-red, to deep-burgundy colour when they first emerge in spring and gradually turn deep green.
Flowers Insignificant small, sweetly scented, greenish flowers with white filaments and red anthers form on short, dense, thickset, 3–6 cm terminal spikes; spikes are simple or have one or two branches.
Fruit Thin, woody, flat pods up to 16cm long are usually hidden in the foliage. Green, yellow, red-brown to yellowish-green at first, they turn grey or brown and smooth when mature. They have a narrow base, broadening towards the apex. Strongly beaked tips split explosively to release the round, flat, brown seeds.
Nomenclature The genus name *Brachystegia* is based in Greek words meaning 'short' or 'flattened', and 'roof' or 'cover', but the allusion is not clear. The specific name *spiciformis* means 'spike-like' and describes the inflorescence.
Use The wood is widely used for fuel, both as charcoal and firewood, and for making beehives; it is considered too inferior for general use.

In the Zimbabwean spring, when the sky is hazy with dust and the smoke of bush fires, the Msasa trees burst into new leaf burnishing the countryside in shades of copper and bronze. The new pale-pink, rusty-orange, red, and deep-burgundy leaves make a woodland full of Msasa trees partcularly glorious, especially when the sun shines through the branches.

So, what makes the spring leaves on the Msasa tree red – a colour traditionally associated with autumn – you might ask? The answer is that the tender new leaves contain a pigment called anthocyanin, which not only gives them their colouring, but importantly acts as a sun screen, preventing sunburn. As the leaves mature and harden, they no longer need the protection and change to green.

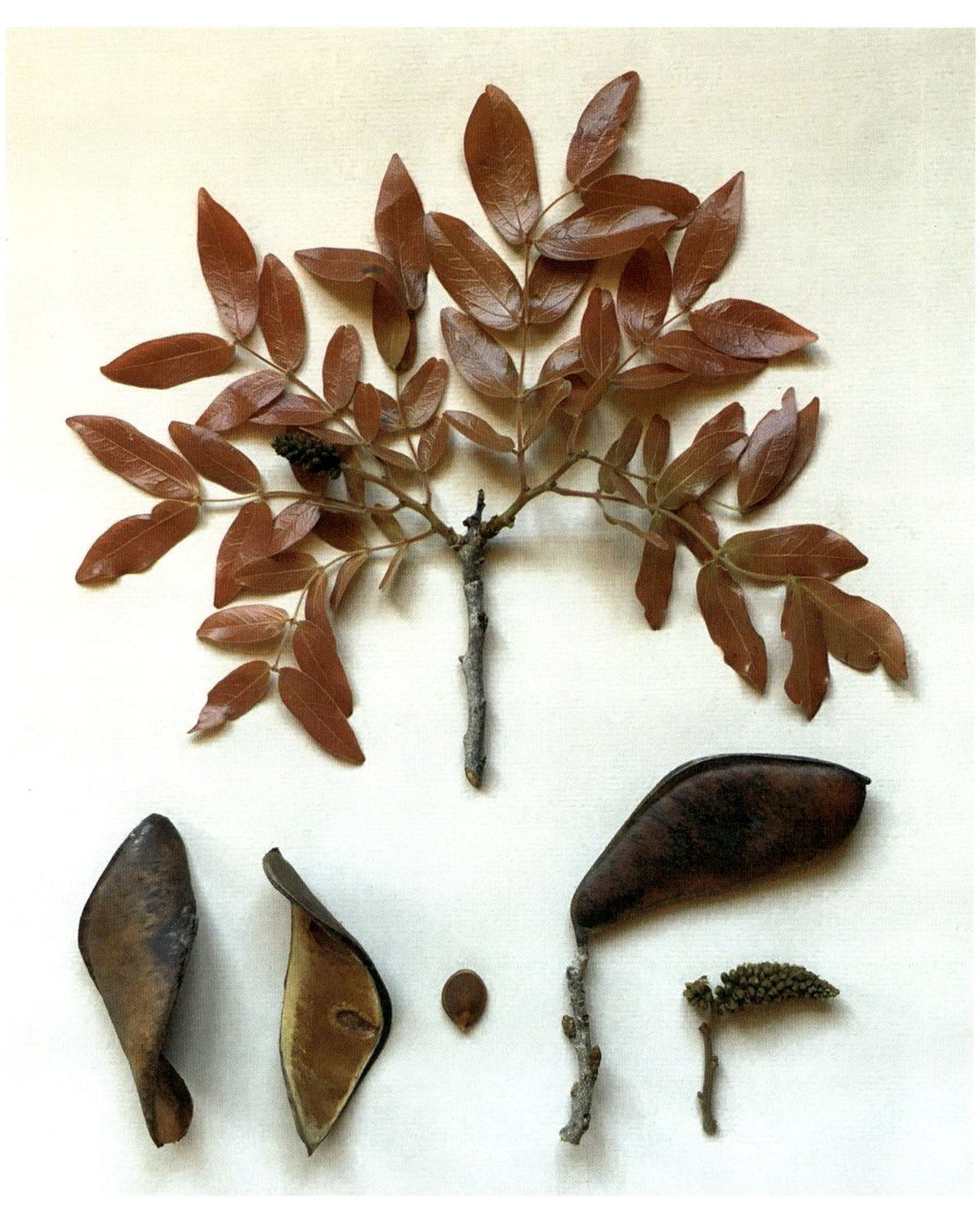

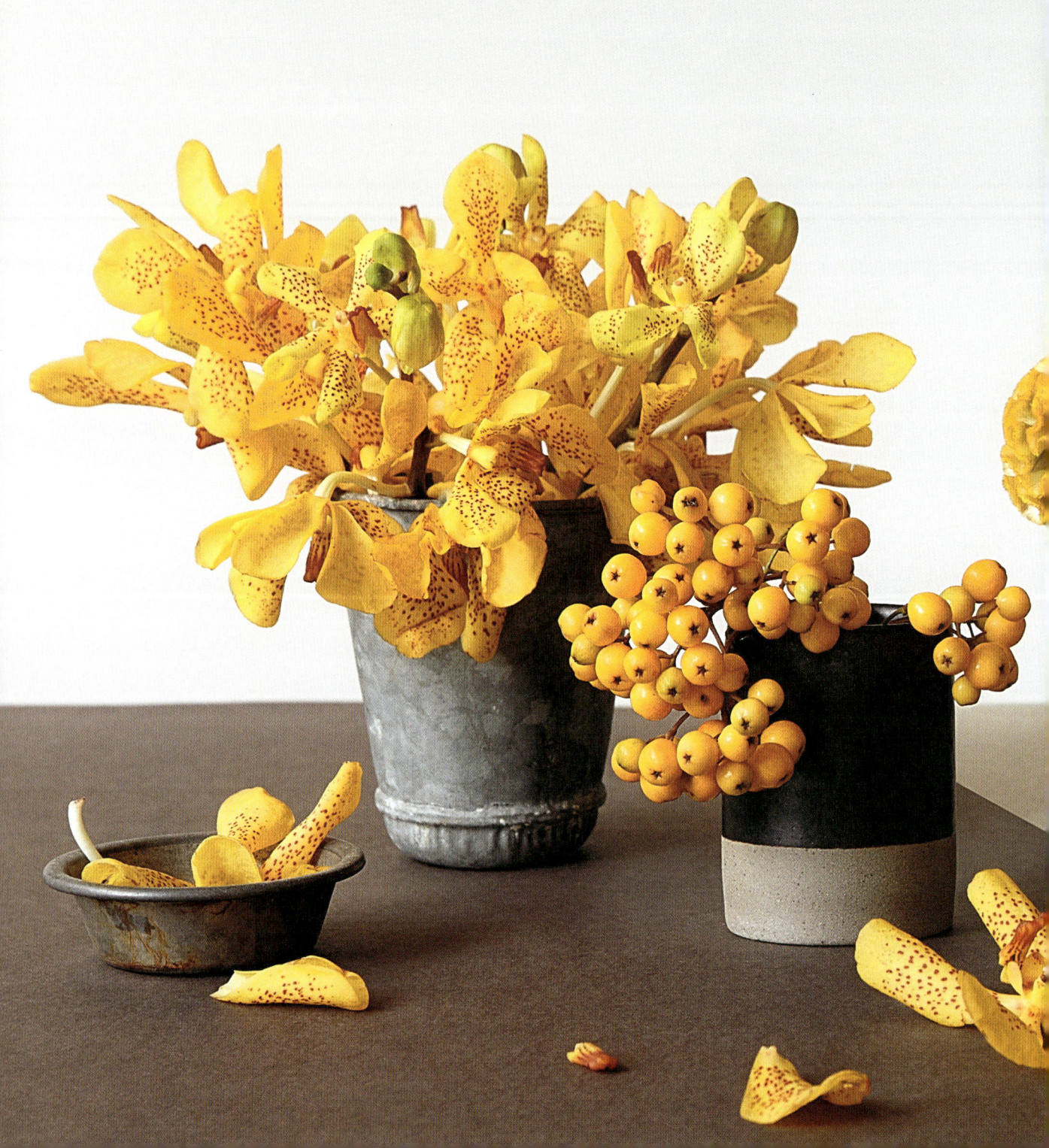

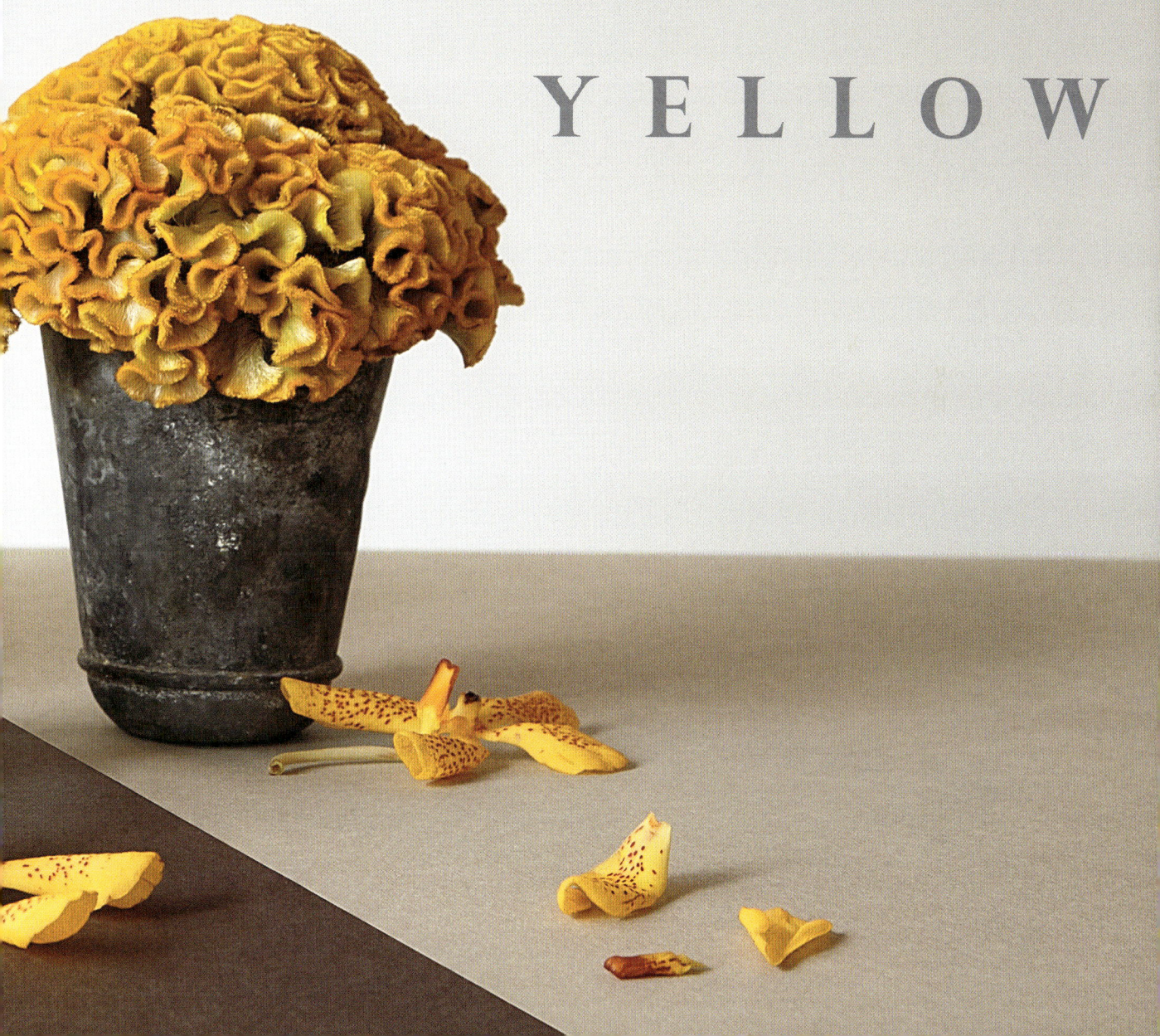

Chapter 3

YELLOW

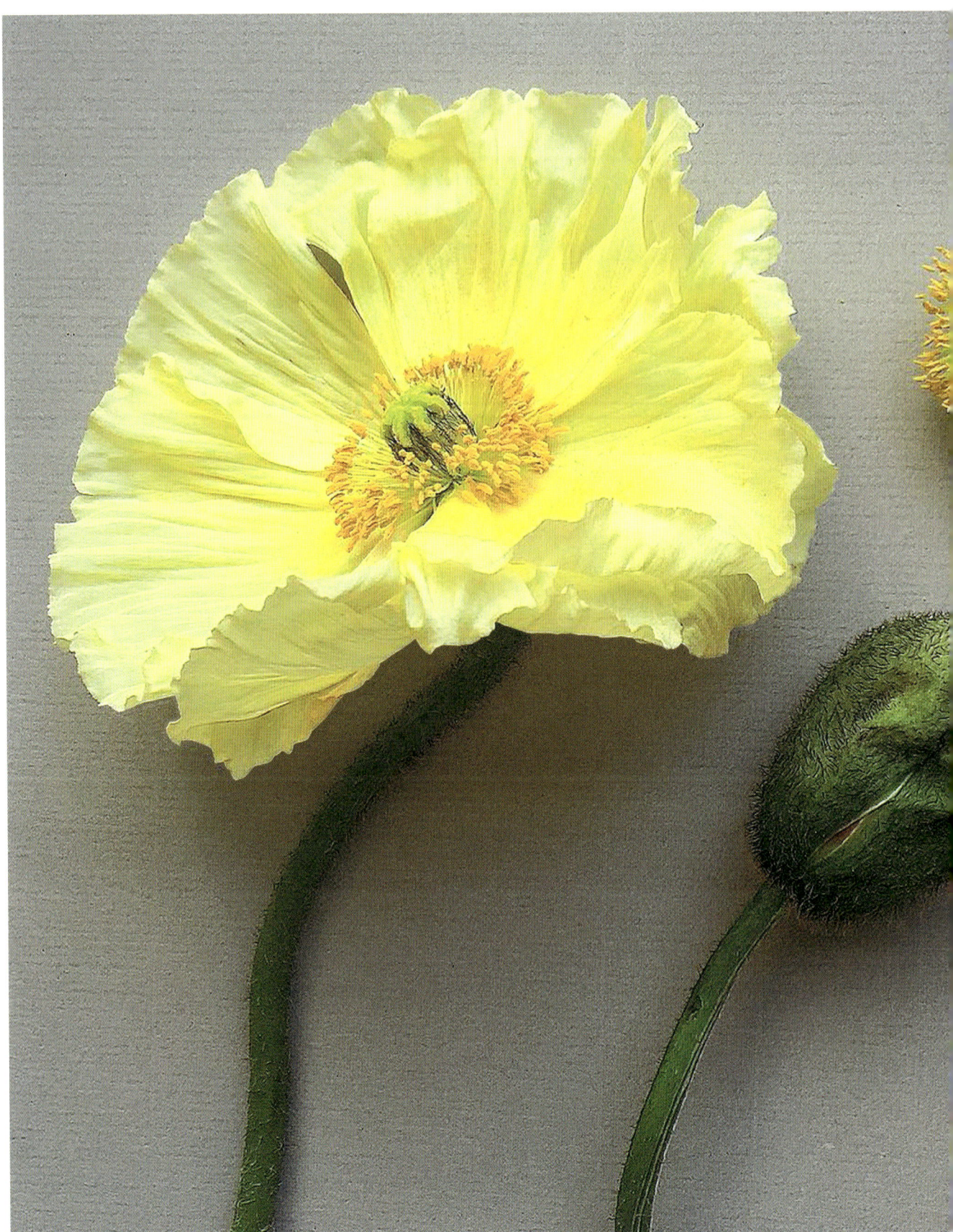

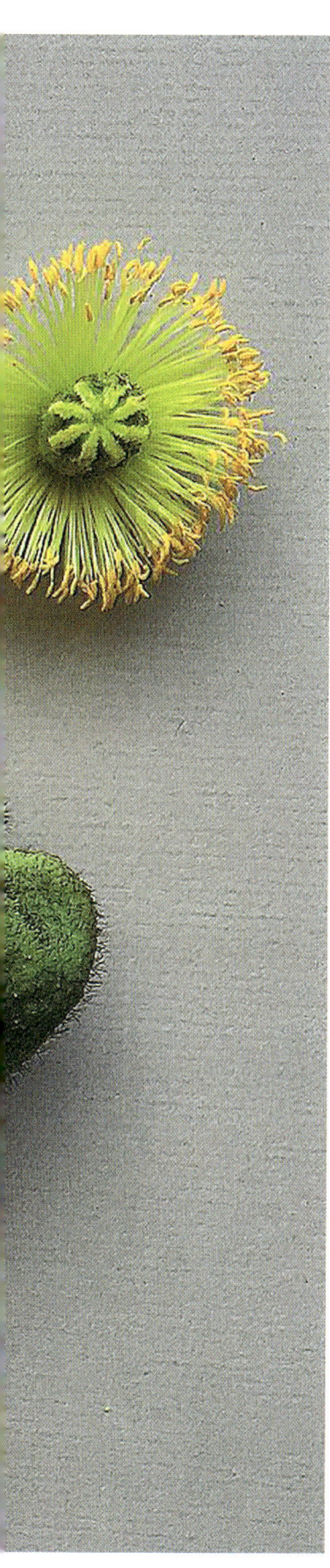

ICELAND POPPY

Latin name *Papaver nudicaule*
Photograph shows Flower; stigma, style, ovary and stamens; bud.
Collected London, UK, 2016
Other common name Arctic poppy
Family *Papaveraceae*
Origin North America, extending into the arctic regions of Alaska, and the mountains of central Asia.
Description Clump-forming, summer-flowering (in its natural habitat), perennial, although it is more commonly grown as annual.
Leaves Hairy, narrow, blue-green leaves grow up to 15cm long and exude a milky fluid (latex) when cut.
Flowers Bears saucer-shaped, mildly fragrant flowers that can be up to 10cm in diameter and have four crêpe-paper textured petals. *P. nudicaule* can be yellow, orange, white, coral, and pink, depending on the cultivar.
Fruit Produces oblong, encapsulated fruit that can be 12–20cm long.
Nomenclature The plant contains a milky latex and, in reference to this, Swedish botanist Carl Linnaeus (1707–78) named it *Papaver* from the Latin word *pappa* meaning 'milk'. *Nudicaule* is from the Latin *nudus* or 'naked', describing the leafless stems.
Uses Grown as a decorative ornamental. The leaves can be cooked and eaten, but the flowers are mildly toxic to mammals. A natural dye can be obtained from the flowers and buds.

Iceland poppies (along with sweet peas, see page 96) were the quintessential late winter annuals grown in Zimbabwe. The crinkled 'crêpe-paper' petals of this flower unfurl from furry 'pixie hat' casings. The combined scent of the blooms and that of the milky stem sap is heady and narcotic. If cut in bud, and the stems singed with a flame to seal them, the Iceland poppy lasts well as a cut flower.

VENICE MALLOW

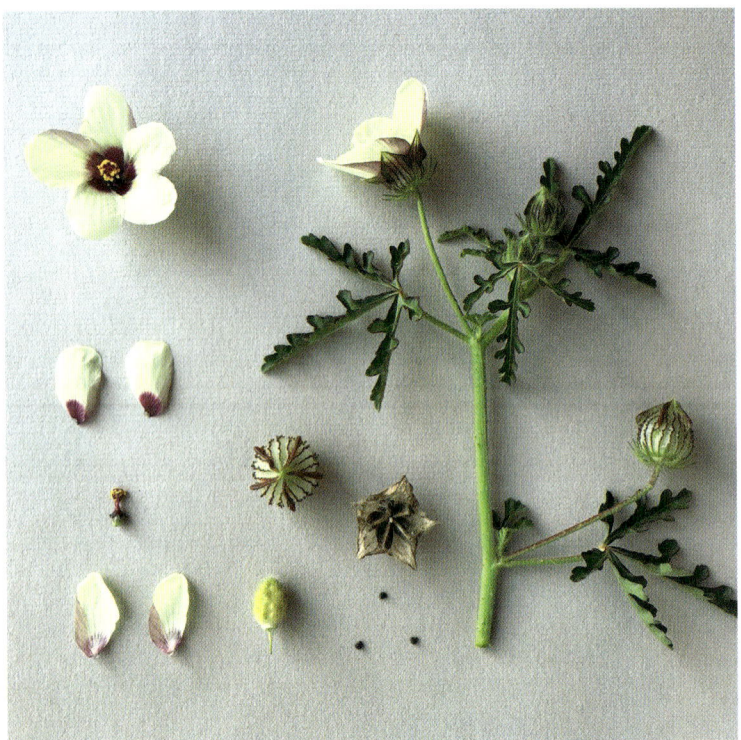

Latin name *Hibiscus trionum*
Photograph shows Flower, section of upper stem with flower and bud, seeds, open seed capsule, immature seed capsule, bud, petals, pistil with ovary and stamens, petals.
Collected Johannesburg, South Africa, 2016
Other common names Flower of an hour, Black-eyed Susan
Family *Malvaceae*
Origin Uncertain, but now widely distributed in southeastern Europe.
Description Fast-growing annual that can grow to height and spread of 60cm.
Leaves These grow to about 7cm. Upper leaves are shallowly divided into two to five lobes, while lower leaves are circular, with coarsely toothed margins.

Flowers The creamy yellow flowers have five sepals with purple bases, and can be up to 5cm in diameter. Actinomorphic and hermaphrodite, they have several stamens with golden anthers, and a style that is divided into five stigmata. Flowers only last for a few days.
Fruit Produces dark-brown, ovoid, hairy capsules up to 3cm across, containing kidney-shaped seeds. The calices swell as the pods mature, then split open to reveal five compartments.
Nomenclature *Hibiscus* is from the Greek *hibiskos* refering to another plant in the *malvaceae* family – the marshmallow. *Trionium* is three-part leaf.
Uses Medicinal – the flowers are used as a diuretic. The edible tender shoots are eaten both raw or cooked.

Once while in transit in Johannesburg I stayed overnight in a lodge on the edge of a dam. Hundreds of these lovely little yellow mallows flowered on some disturbed ground nearby.

WATER LILY

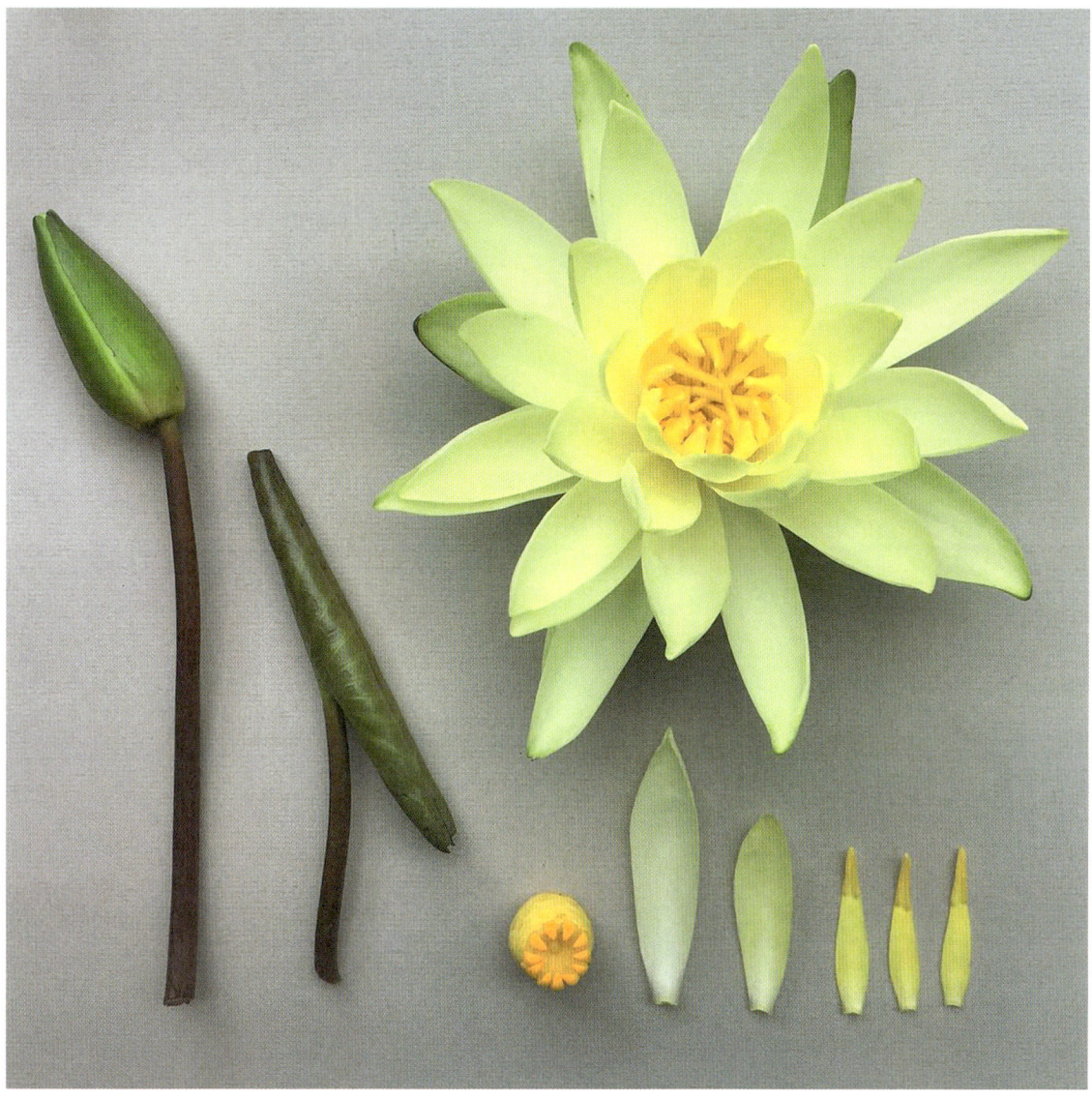

Latin name *Nuphar lutea*
(syn. *Nymphaea lutea*)
Photograph shows Bud, unfurled leaf, flower, anthers, petals, pistil.
Collected Harare, Zimbabwe, 2015
Family *Nymphaeaceae*
Origin India
Description Aquatic plant that flowers all year round; leaves and flowers generally float on the water's surface.
Leaves Peltate, circular leaves have sharp-toothed margins and can be up to 40cm in diameter.
Flowers Bears intensely yellow flowers, 5–6cm across, that consist of four sepals and many petals. They open at dawn and close at dusk.
Fruit Produces greenish-brown, round berries, which contain numerous globular seeds.
Nomenclature The genus name *nymphaea* was inspired by the nymphs of Greek mythology.
Uses As well as being used for ornamental purposes, water lilies are important in the ecosystem of freshwater lakes and ponds, providing food for herbivorous animals and habitat for others. They are also a food source to local communities in India.

I had to wade barefoot into the dam on a friend's property to collect this specimen. At Maduma, we have a pond filled with yellow water lilies that my grandmother used pick to decorate the dining table. She would pour hot wax into the stamens to stop the flowers closing, then float them on water in small glass bowls.

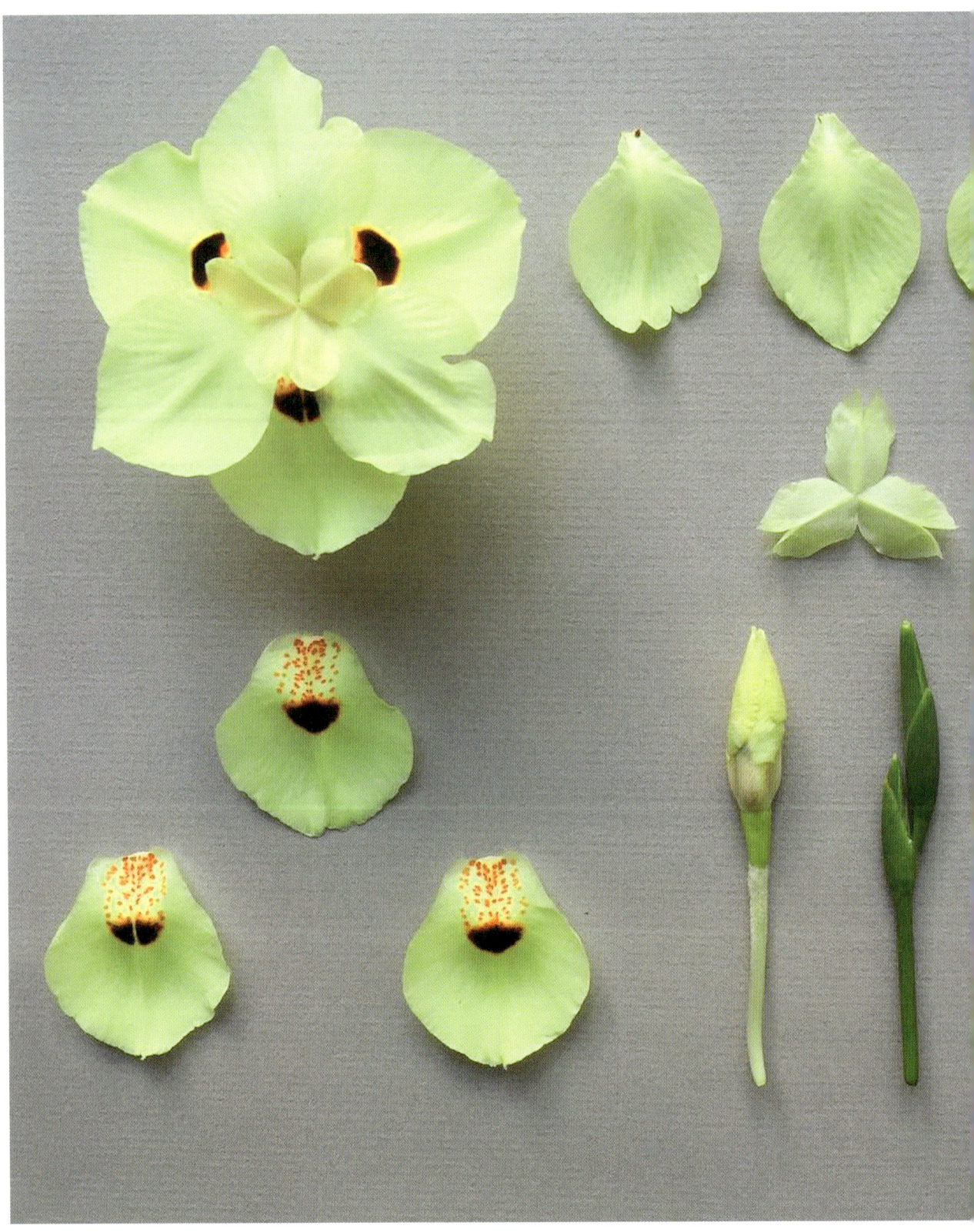

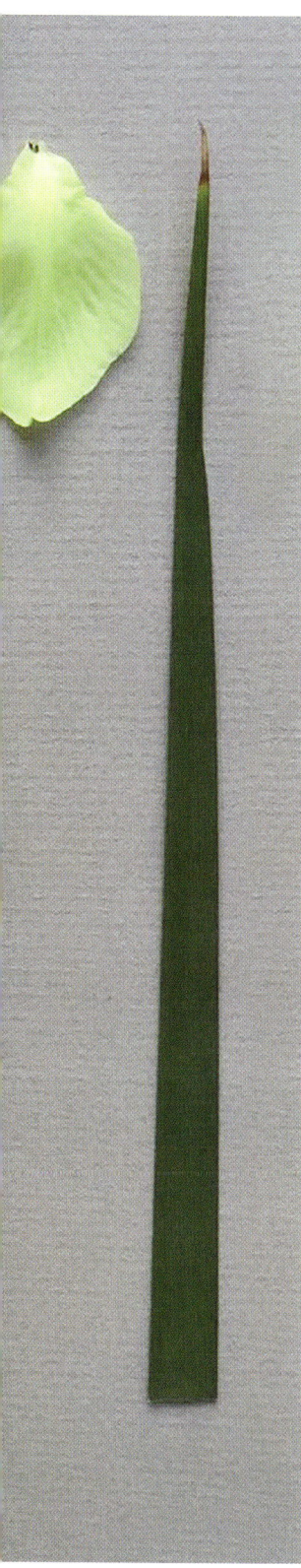

AFRICAN IRIS

Latin name *Dietes bicolor*
Photograph shows Flower, staminodes, leaf, immature bud, bud, style arms, tepals.
Collected Long Valley, Harare, Zimbabwe, 2016
Other common name Fortnight lily, Wild iris
Family *Iridaceae*
Origin South Africa
Description Evergreen perennials grown from rhizomes that form clumps of erect sword-shaped leaves arranged in flat fans similar to other members of the iris family. The adult plant is approximately 1m wide and 1m tall.
Leaves These are light green and laneolate, around 1–2cm wide, and have a double central vein.
Flowers Bears pale-yellow flowers, 5cm in diameter, from spring to summer.

Blooms last only one day, but are quickly replaced. Each one has three light-yellow tepals with dark-brown blotches at the bases, and three petal-like staminodes without blotches. Flowering occurs in bursts at two-week intervals.
Fruit Produces club-shaped capsules approximately 25mm in diameter that partially open to release seeds.
Nomenclature Genus name comes from a commonly used prefix *di* or *dis*, meaning 'double' and *etes*, or 'an associate' referring to the position of this genus between *Moraea* and *Iris*. The specific epithet *bicolor*, means having two colours.
Uses Species is drought tolerant, low maintenance, and has fire retardant properties. In southern Africa, the roots were traditionally worn to protect and strengthen the wearer.

These pale-yellow flowers with their dark blotches have always reminded me of the delicious Portuguese custard tarts, *pastel de natas*, that I have a penchant for.

GWANGWADIZA

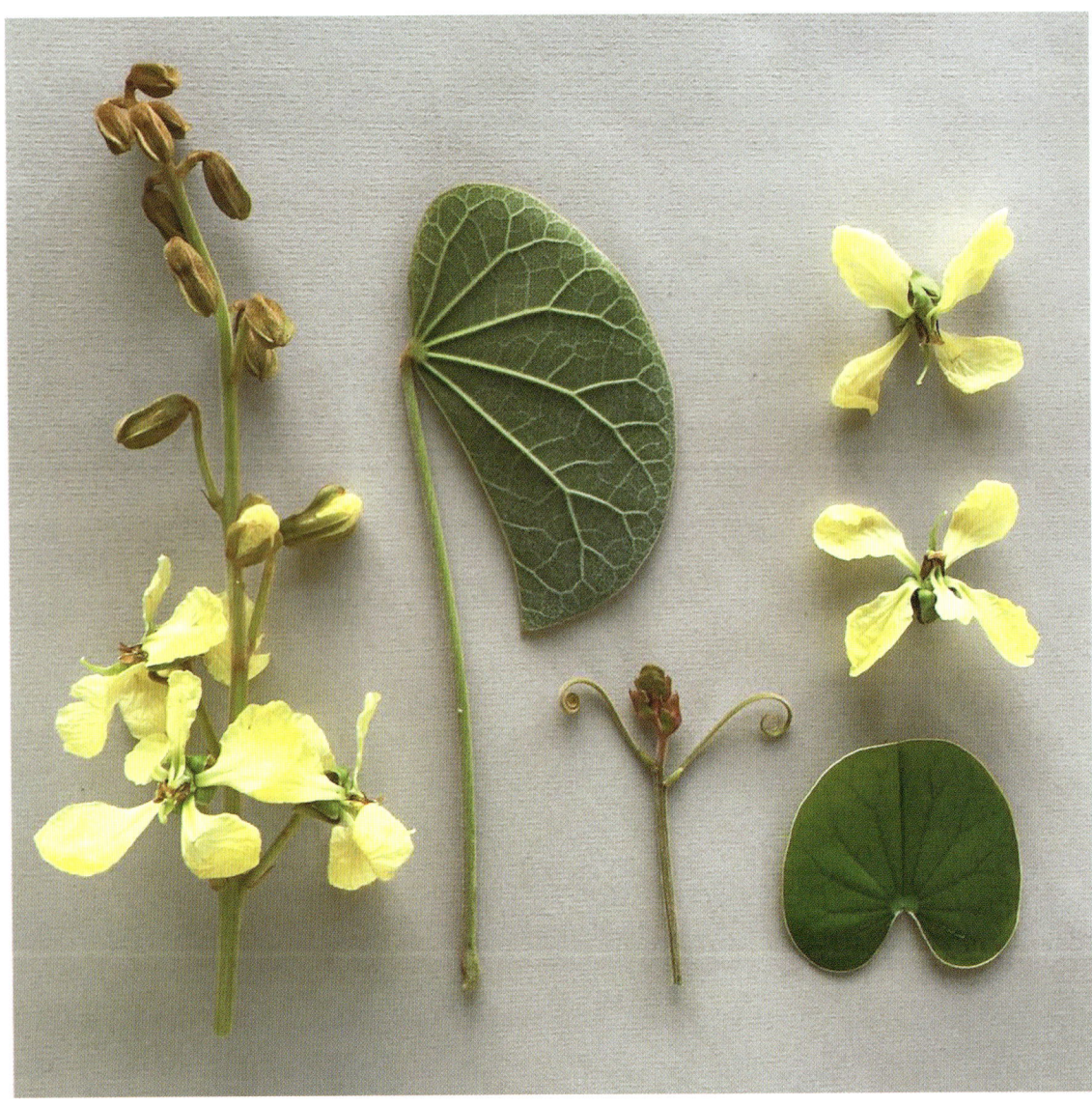

Latin name *Tylosema fassoglensis* (syn. *Bauhinia fassoglense*)

Photograph shows Inflorescence, closed leaf, flowers, open leaf, immature leaf with tendrils.

Collected Mutare, Zimbabwe, 2015

Other common names Creeping Bauhinia, Marama bean – *Gwangwandiza* is its Shona name.

Family *Fabaceae*

Origin Sudan, Ethiopia, South Africa at elevations of up to 2,100m, in grasslands and woodlands.

Description A large, tuberous perennial creeper that grows up to 6m.

Leaves Alternate, bilobate leaves are up to 20cm long. They are usually hairy (pubescent) on the underside with tendrils that attach to lateral branches.

Flowers Bears inflorescences on racemes 5-45cm long. Small zygomorphic, bisexual flowers have five petals; two are united and three are free.

Fruit Forms flat, woody, oblong seed pods, up to 7.5cm long, which contain one or two brown to black ellipsoid seeds.

Nomenclature Species names comes from the Greek words *tylos* meaning 'knob' and *sema* meaning 'sign', referring to the large seeds of the plant. *Fassoglensis* comes from Fazoghli, a place in Sudan, where the species was first encountered.

Uses Seeds can be eaten raw or cooked. The fibre is suitable for cloth. In Kenya people make rope and plaited items from the stems or their fibres. Tubers and roots are eaten and used in traditional African medicine.

This scrambling creeper with its radiating stems and lovely yellow flower grows extensively on the roadside verges in Zimbabwe.

OBSCURE MORNING GLORY

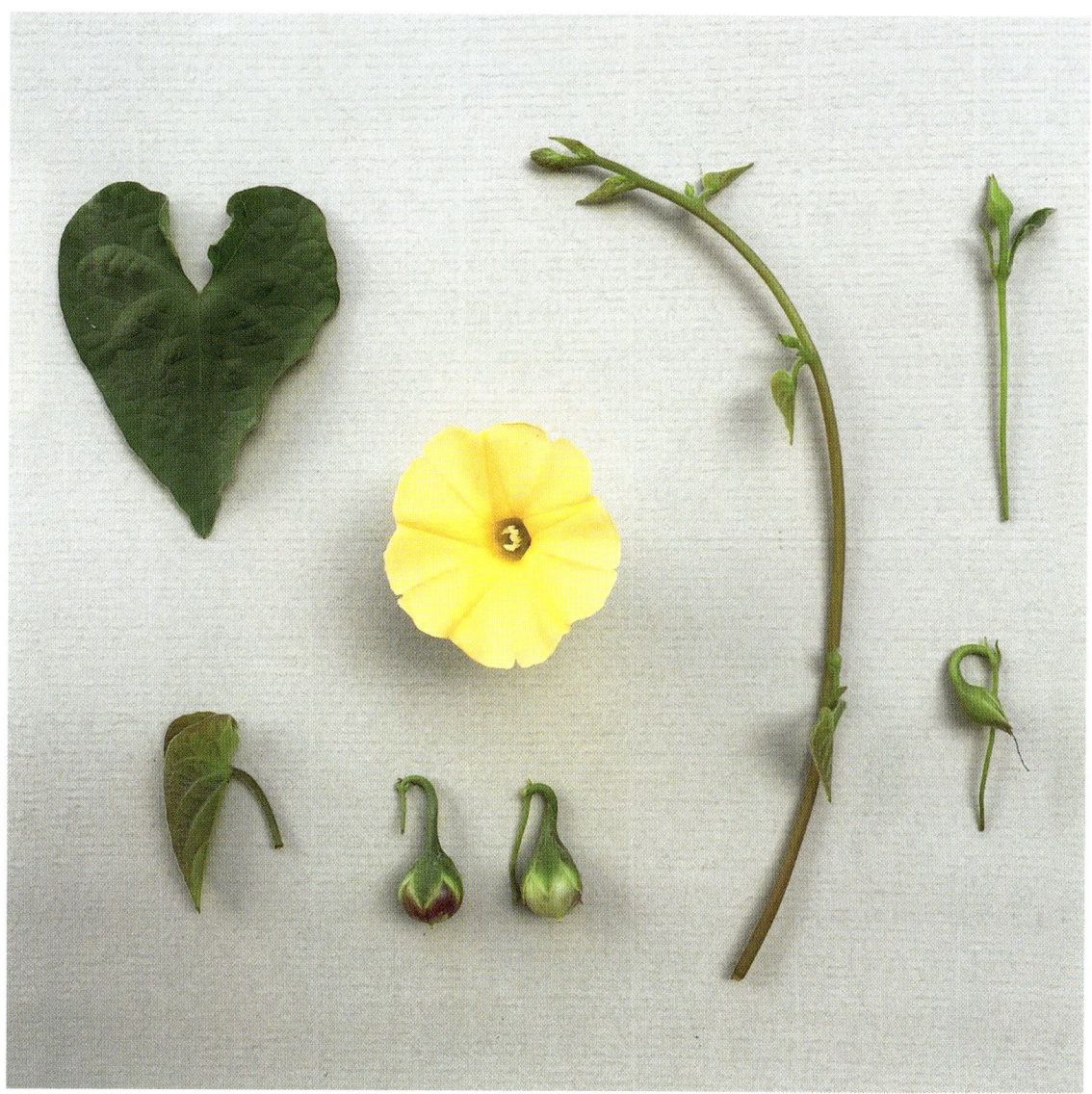

Latin name *Ipomoea obscura*
Photograph shows Mature leaf, flower, tendrils with buds, fruit, young leaf.
Collected Harare, Zimbabwe, 2016
Family *Convolvulaceae*
Origin Tropical Africa and Australia, Asia, islands of the South Pacific.
Description Annual or perennial twining vine that grows up to 3m from a taproot.
Leaves Green, alternate, heart-shaped leaves have entire margins and can be 2.5–8cm long.

Flowers Trumpet-shaped, pale-yellow flowers are 1–2cm across, and have maroon centres.
Fruit Forms egg-shaped, light-brown capsules with a sharp, pointed apex.
Nomenclature Species name is from the Greek words *ips* and *homoios*, meaning 'worm' and 'like' respectively, and *obscura* means 'obscure' or 'indistinct'.
Uses The leaves can be cooked, and have a pleasant taste. The sap, leaves, and roots are used in traditional medicines.

I collected this pretty, pale-yellow *Ipomoea* in our church car park one Sunday after the service.

LEOPARD ORCHID

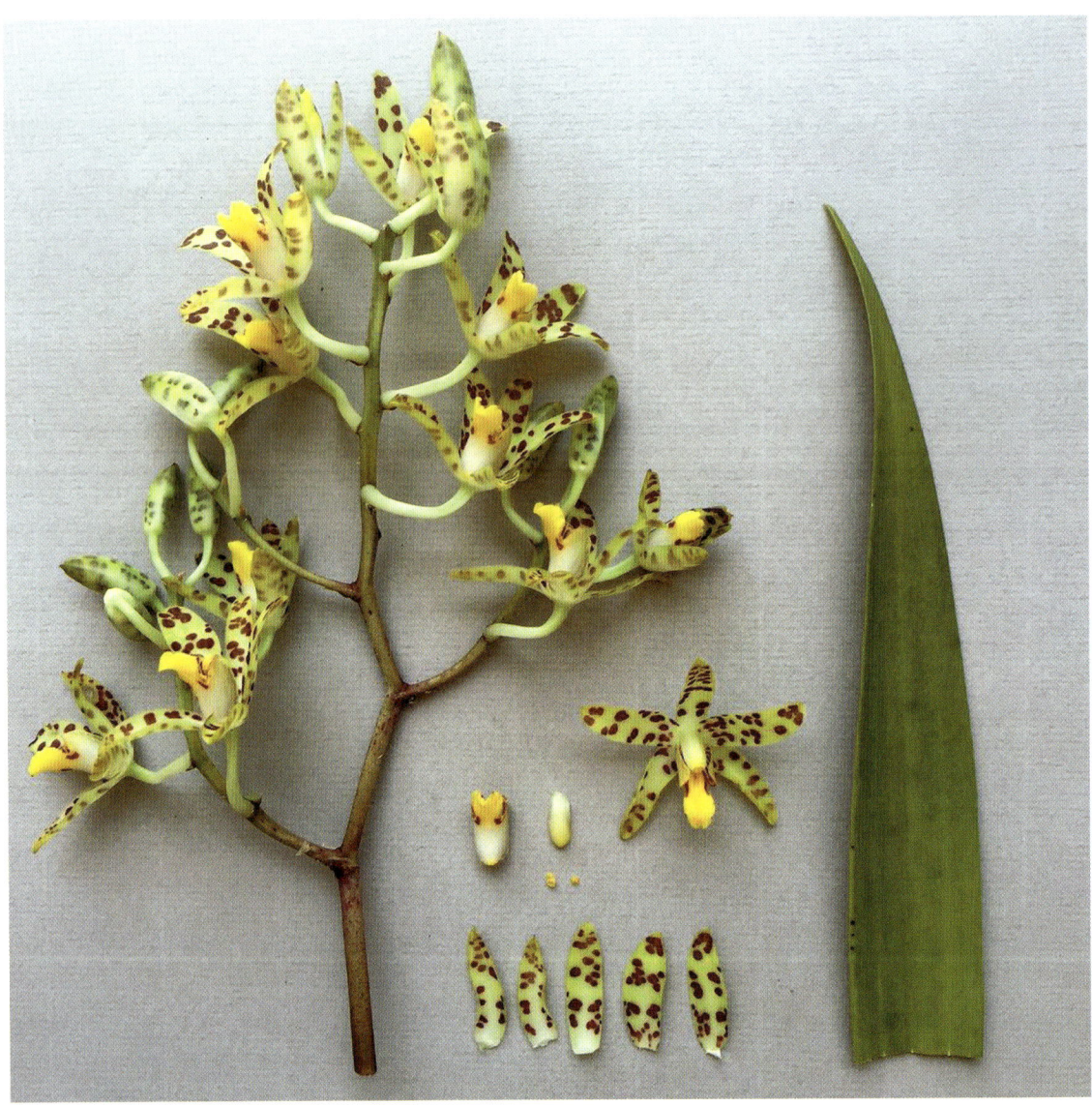

Latin name *Ansellia africana*
Photograph shows Flower spike, leaf, sepals, petals, labellum, column, anther cap, pollinia, single flower.
Collected Ewanrigg Botanical Gardens, Harare, 2015
Family *Orchidaceae*
Origin Southern Africa – Namibia, Botswana, Swaziland.
Description The largest of the epiphytic orchids in southern Africa, *A. africana* can be up to 80 cm tall and is monotypic. It grows as clumps on trees and has upwards pointing roots that form a dense mat and anchor plant to the tree.
Leaves Long, narrow leaves grow in

between the pseudo-bulbs.
Flowers Terminal inflorescence bears many individual, large, showy, fragrant flowers. They are variable, usually yellow, and can have maroon or brown spots. They are pollinated by moths.
Fruit Forms an oblong green, capsule.
Nomenclature Species is named after the British botanist and traveller John Ansell (died 1847) and then *africana* from its African origins.
Uses It is much in demand as an ornamental. Traditionally it is used as a love charm and an antidote to bad dreams. Tea can be made with the pseudo-bulbs.

The was the first indigenous tree orchid I encountered. Single specimens of *Ansellia* can reach an incredible size and eagle owls have been known to nest in clumps of this epiphyte.

CHALICE VINE

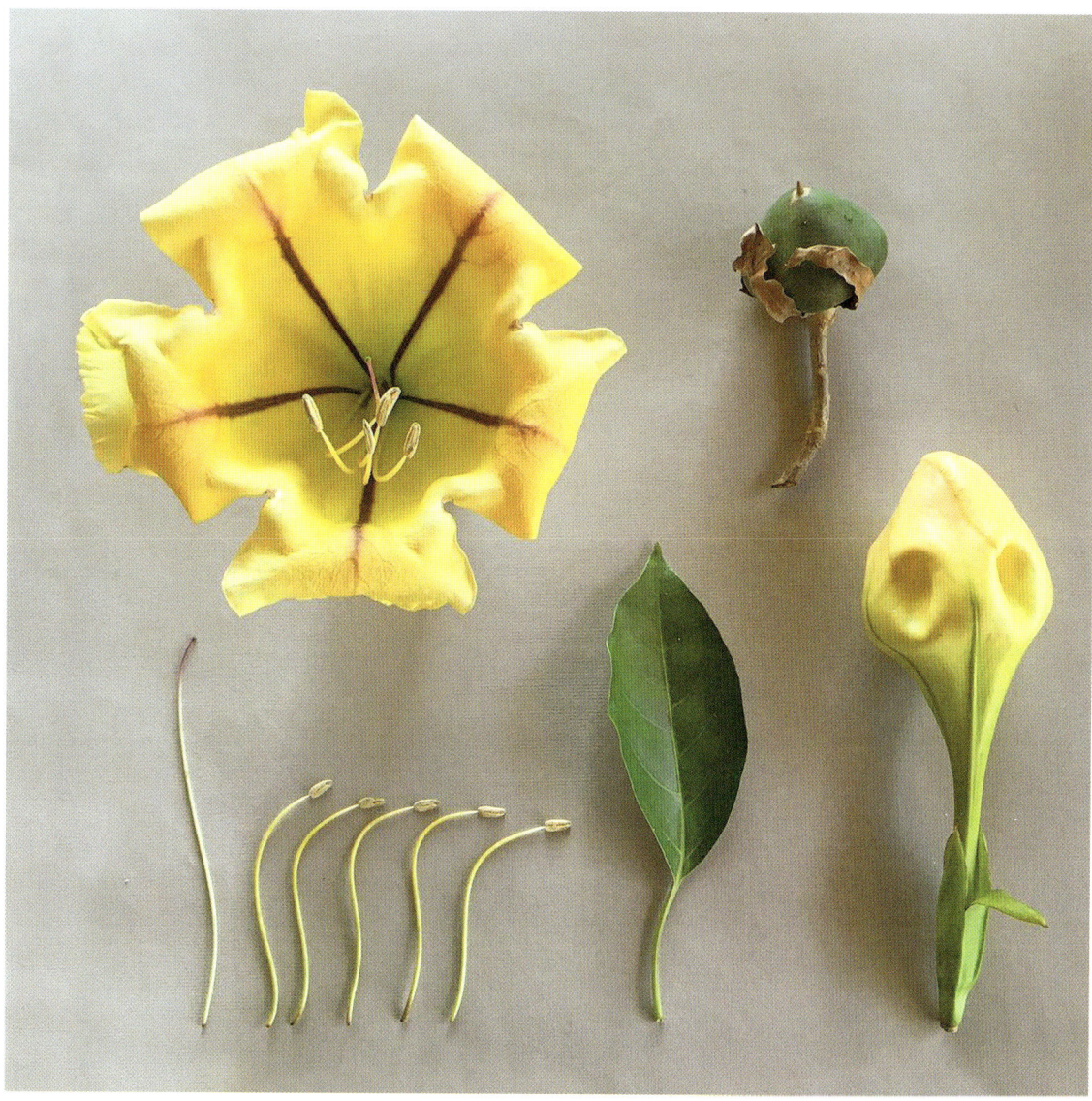

Latin name *Solandra maxima*
Collected Melfort, Zimbabwe, 2016
Photograph shows Flower, seed capsule, bud, leaf, stamens, pistil.
Other common names Cup of gold, Copa de oro
Family *Solanaceae*
Origin Mexico, Central America, Colombia, Venezuela.
Description Evergreen, woody scrambling climber that often extends over 50m, and mainly blooms in the winter dry season.
Leaves Leathery, medium-green, glossy, elliptic leaves are up to 15cm long and have long petioles; emerging leaves are purple, becoming green later.

Flowers Bears dimpled buds that open into chalice-shaped flowers up to 20cm long; the blooms turn from yellow to gold as they age. The flowers, pollinated by bats, are especially fragrant at night and smell like coconut.
Fruit Conical green berries are 5cm in diameter and turn red as they ripen.
Nomenclature The species *Solandra* is named after the Swedish naturalist, Daniel Solander (1733–82), and *maximus*, Latin for 'greatest', describes the very large flowers.
Uses Decorative ornamental. The species is poisonous, but it is used for its hallucinogenic properties in sacred ceremonies by the Mexican Huichol tribe.

One cold November's day in London, UK, I noticed a vine covered with oversize flowers growing up the side of a house in Chelsea. To my amazement it was a tropical *Solandra maxima* not only thriving in London's micro-climate, but also blooming in the depths of winter. The scent of this vine's huge flowers is reminiscent of piña coladas, one of my guilty pleasures.

FRANGIPANI

Latin name *Plumeria alba*
Photograph shows Flower cluster
Collected Harare, Zimbabwe, 2015
Other common names West Indian jasmine, Jasmine mango
Family *Apocynaceae*
Origin Puerto Rico and Lesser Antilles.
Description Medium-sized, evergreen tree with an open-vase form that grows up to 8m tall and flowers all year round.
Leaves Thick, leathery leaves up to 30cm long, have pointed tips, and are spirally arranged at the end of branches.
Flowers Showy, fragrant, tubular, flowers, each up to 5cm across, are borne on terminal inflorescences. They vary in colour from white, yellow, to pink.

Fruit Cylindrical seed pods up to 12cm long turn from green to brown with age; each contains up to 60 winged seeds.
Nomenclature The species *Plumeria* was named after the 17th-century French botanist Charles Plumier (1664–1704). The specific epithet is derived from the Latin *alba* meaning white. The common name 'Frangipani' is taken from an almond-scented, glove perfume used in the 16th century, which the flower's scent is reminiscent of.
Uses In Hawaii it is grown for its flowers – around 14 million blooms are used each year to make *leis*, or garlands. The tree has medicinal properties; the flowers are used in traditional Chinese medicine.

The flowers of this tree have a wonderful warm, almondy scent. My grandmother planted an avenue of them on the back drive at Maduma. On one of the trees grew a large epiphytic Staghorn fern — named for their antler-shaped fronds. The ferns basal fronds resemble shields and grow in overlapping layers. Staghorns thrive on potassium and my grandmother would tuck banana skins behind its papery shields.

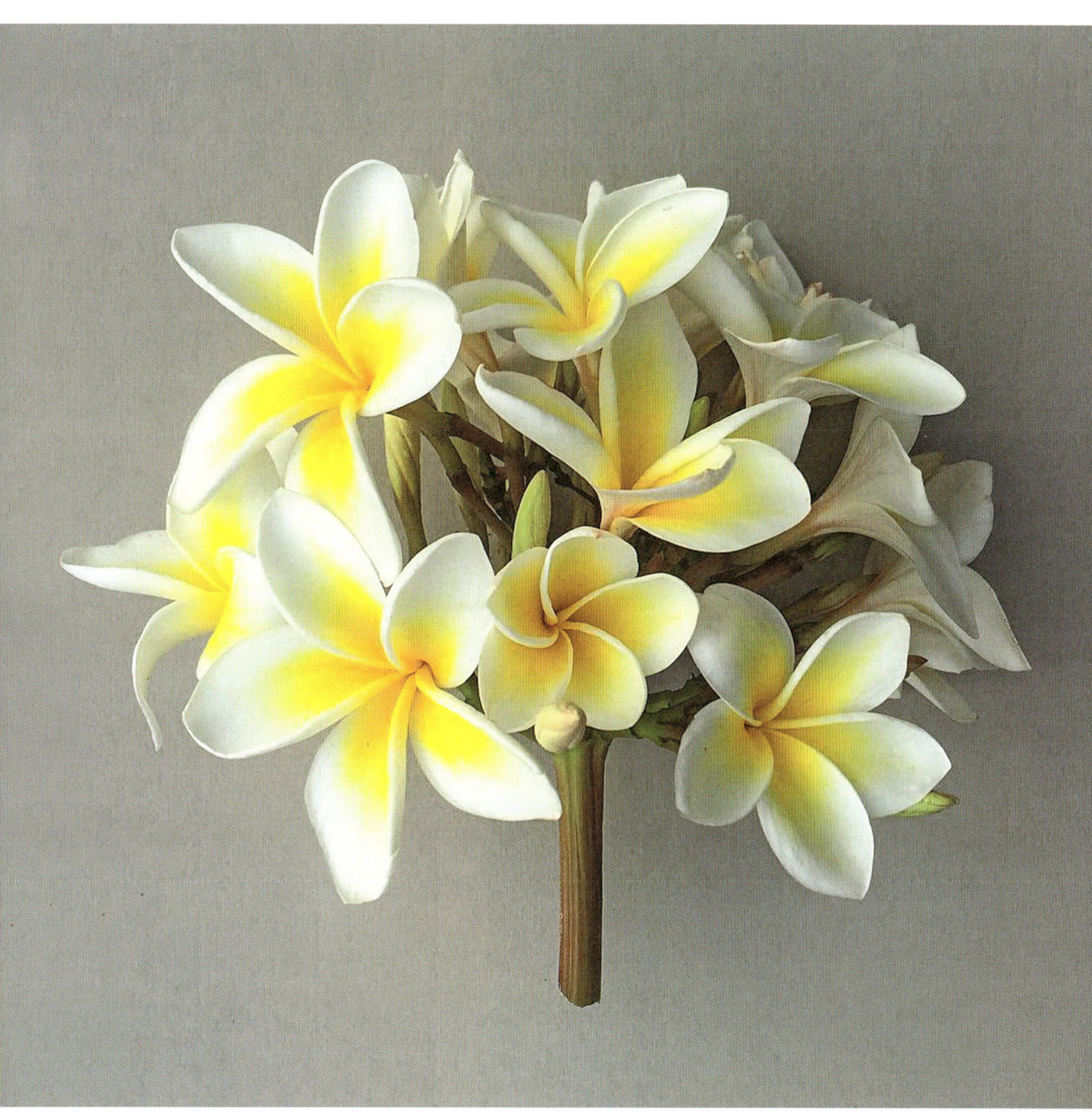

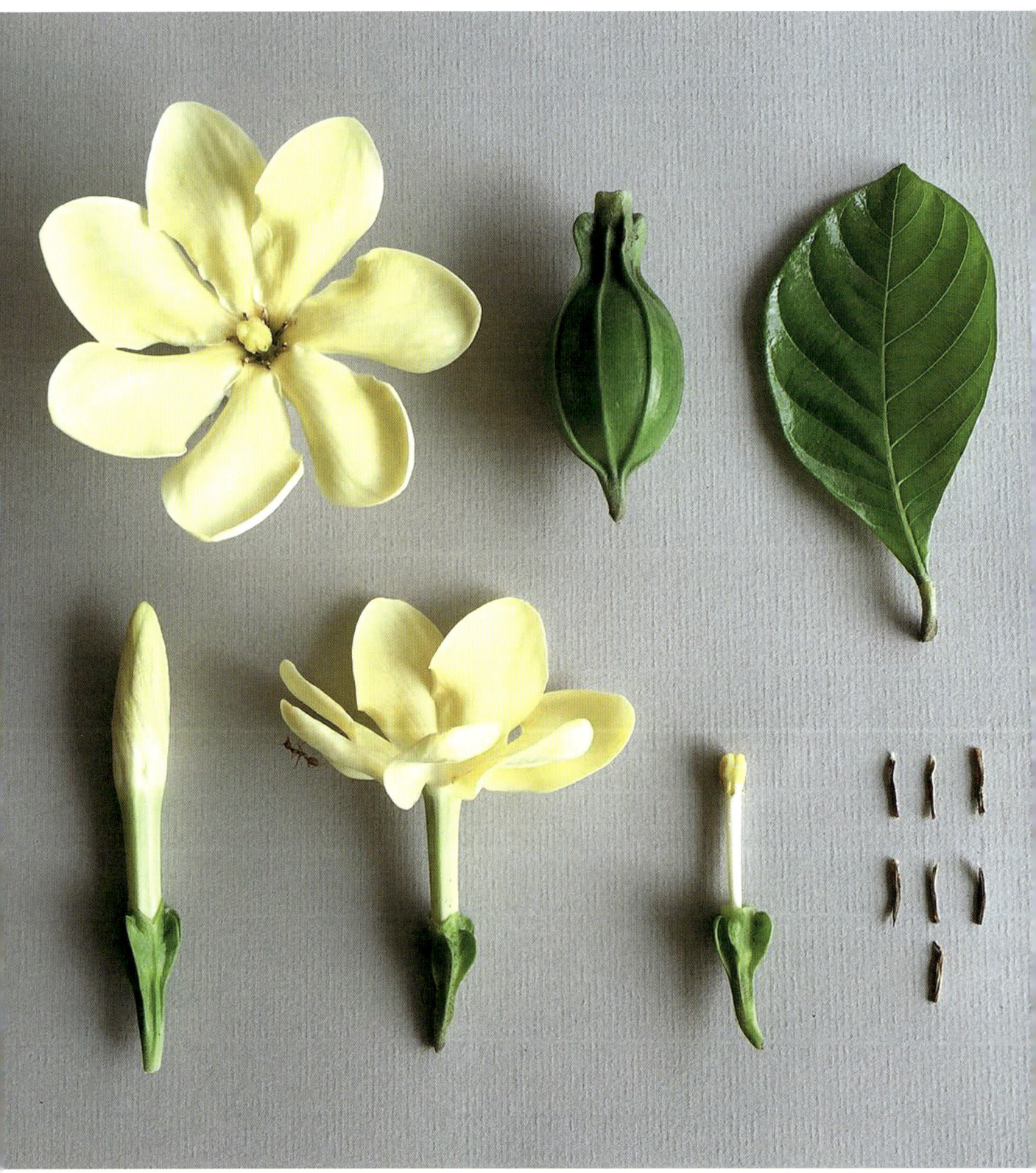

KEDAH GARDENIA

Latin name *Gardenia carinata*
Photograph (opposite) shows Lighter yellow flower, pod, leaf, stamens, pistil, side view of flower with a red weaver ant (*Oecophylla smaragdina*), bud.
Collected Kuala Lumpur, Malaysia, 2016
Other common name Golden gardenia
Family *Rubiaceae*
Origin Malaysia
Description A small evergreen tree or large shrub that grows up to 3m in height.
Leaves The elliptic to obovate leaves have prominent veins, entire margins, and hairy undersides.

Flowers Fragrant, pinwheel-shaped flowers (11.4cm in diameter) are white initially, but darken to yellow, then orange over the course of three days.
Fruit Produces round to ovoid pods with longitudinal ridges. They are yellowish-green at first, then turn brown to black as they dry.
Nomenclature Species is named after the Scottish-born, American botanist Dr Alexander Garden (1730–91). The specific epithet *carinata* comes from a Latin word meaning 'keel' and refers to the plants keel-shaped leaves.
Uses Decorative ornamental.

Leaving Kuala Lumpur airport, *en route* to Langkawi Peninsula, I spotted this gardenia growing by the side of the road. Always up for identifying a new plant I scrambled into the undergrowth to take a closer look, but in doing so upset a nest of weaver ants and within seconds I was being nipped in places no-one wants to be bitten. However, the welts inflicted were a small price to pay for the beautiful flowers I collected.

This photograph, right, shows the darker flowers from the same plant. Seen here is a flower head from above, dried pod, seed, open capsule shell, petals, side view of flower, pistil, and stamens.

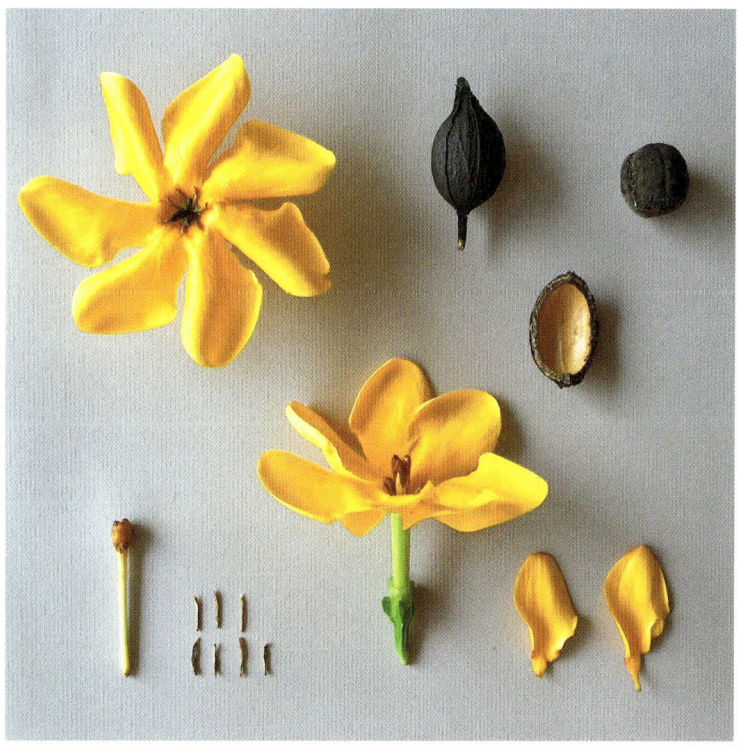

NATAL MAHOGANY

Latin name *Trichilia emetica* sub sp. *emetica*

Photograph shows Flowers, buds, immature fruit, leaf, seeds, ripened fruit.

Collected Maduma, Harare, Zimbabwe, 2015

Other common names Mafura butter tree, Ethiopian mahogany tree

Family Meliaceae

Origin Southern Africa, from South Africa's KwaZulu-Natal Province to tropical Africa. There are two sub species: *T. emetica* sub sp. *emetica* is restricted to southern Africa; *T. emetica* sub sp. *superosa* occurs north of Zambezi river.

Description Evergreen, medium to large tree that can be up to 30m in height. It has a dense, spreading crown. Trees are male or female.

Leaves Alternate, imparipinnate compound leaves have three to six pairs of leaflets and a single terminal one. Leaflets are dark glossy green with rounded or broadly pointed tips. The lower surface is sparsely to densely hairy, with 11–18 closely spaced pairs of principal side veins.

Flowers Small yellowy green, sweetly scented flowers are produced on short axillary panicles in winter to early spring (dry season). Each has five thick petals about 2cm long, that surround hairy central stamens.

Fruit Dehiscent capsules are 18–25mm in diameter and have a 5–10mm neck. Each one contains three black seeds that are almost completely enveloped by a bright red aril.

Nomenclature The genus name *Trichilia* comes from the Greek word *tricho* meaning 'three parts', referring to the fruit's seeds. *Emetica* derived from 'emetic' – a substance used medicinally to induce vomiting.

Uses In southern Africa, the light-brown wood is used extensively for carving and to make musical instruments. The powdered bark is a powerful emetic, and the seeds soaked in water make a milky soup that is eaten with spinach.

An enormous Natal mahogany tree grew in the garden at my grandparent's first home in Salisbury. When my grandfather built Maduma, he took some seeds from the tree and grew a slew of them, which he then planted extensively in the grounds of his new property – 20 of those original trees still stand today. Although the flowers of the Natal mahogany are somewhat insignificant, they redeem themselves with a wonderful lemony jasmine scent that carries on a breeze. They are followed by fruit, which contains the striking red and black seeds also known as 'lucky beans'.

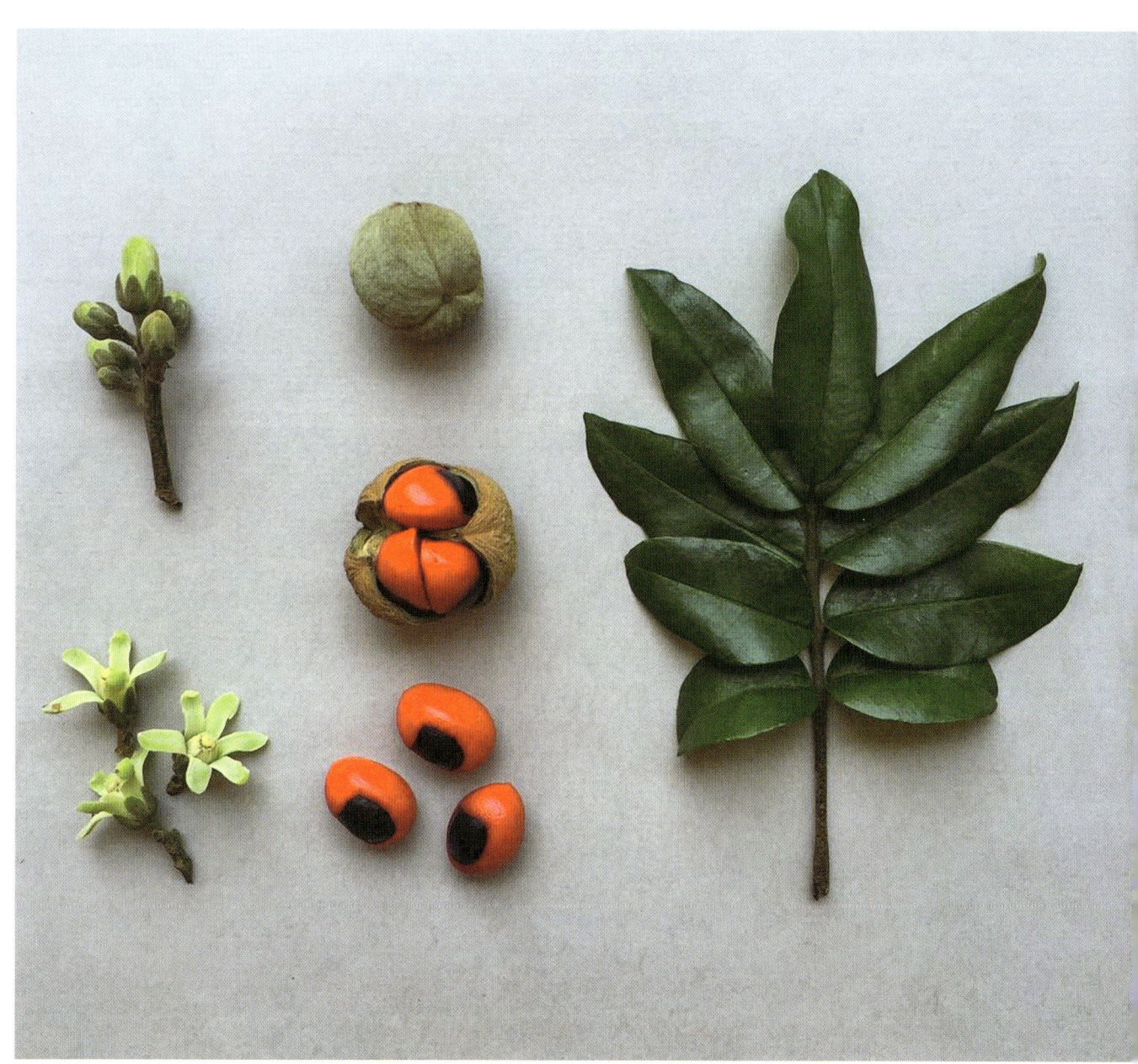

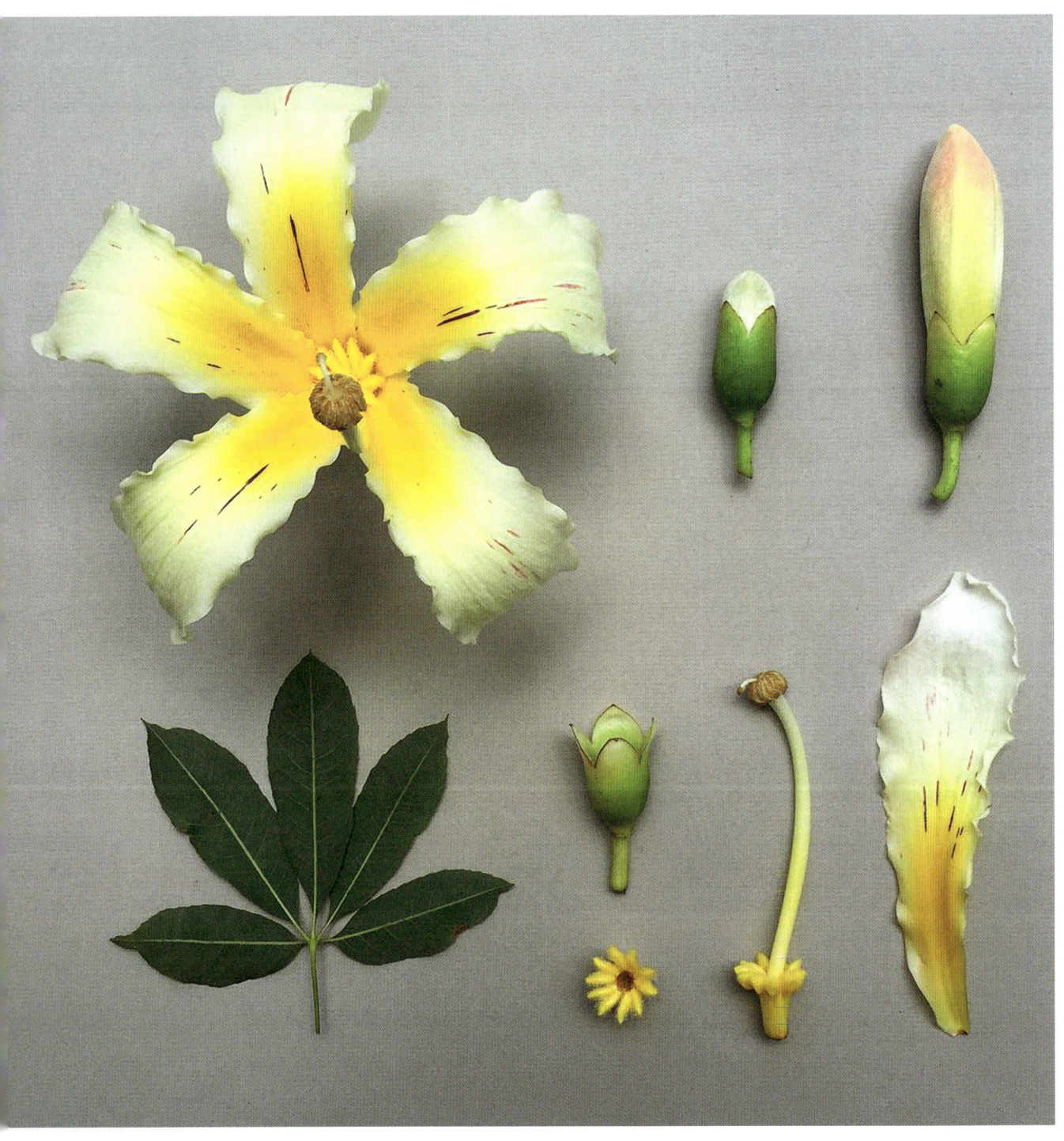

SILK FLOSS TREE

Latin name *Ceiba insignis*
Photograph shows Flower, buds, petal, staminal tube containing stigma, calyx, anthers, leaf.
Collected Harare, Zimbabwe, 2015
Other common names White kapok, Bottle tree, White dragon, Drunken tree
Family *Malvaceae* (formerly *Bombaceae*)
Origin South America – Brazil, Argentina, Paraguay, Peru, Ecuador.
Description A autumn-winter flowering, fast-growing deciduous tree that can be up to 12m in height. It develops a distinctive swollen, bottle-shaped trunk covered in large spines and its bark turns from green to grey as it matures.
Leaves Compound leaves made up of five to seven oblong or obovate, 8-cm leaflets in a palmate arrangement.
Flowers The creamy white or yellow solitary flowers have streaks of pink in the centre and are up to 13cm in diameter. They have five petals and joined filaments that protrude to enclose the stigma. They are pollinated by nocturnal animals.
Fruit Woody, pear-shaped capsules are up to 13cm long and 5–7cm in diameter. They split open when dry, releasing seeds cradled in dense silky floss, hence the common name.
Nomenclature Species name comes language of the Taino people, and means literally 'giant tree'; *insignis*, is from Latin meaning 'remarkable'.
Uses Cloth can be made from fibres obtained from the bark; cushions and mattresses can be stuffed with the floss (kapok) in seed capsules; the soft, light wood is used as a cork substitute.

I collected these flowers from a huge specimen growing (rather inauspiciously) outside a TM supermarket in Harare. The tree has been there all my life and I have long marvelled at its giant Baobab-like shape and large yellow flowers. I recently discovered that the tree had been cut down to make way for a new shopping complex; I am always sad when a botanical hero disappears.

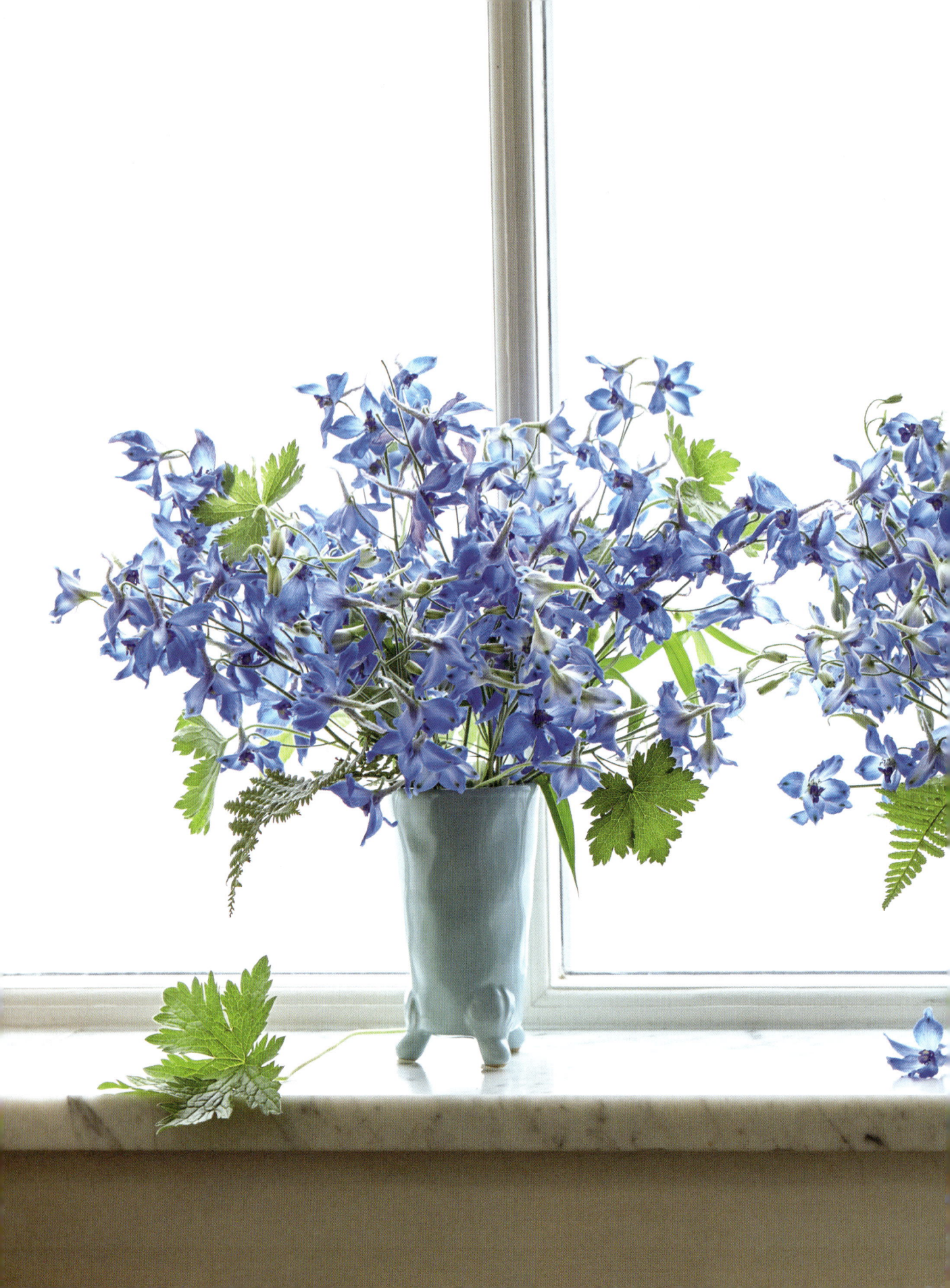

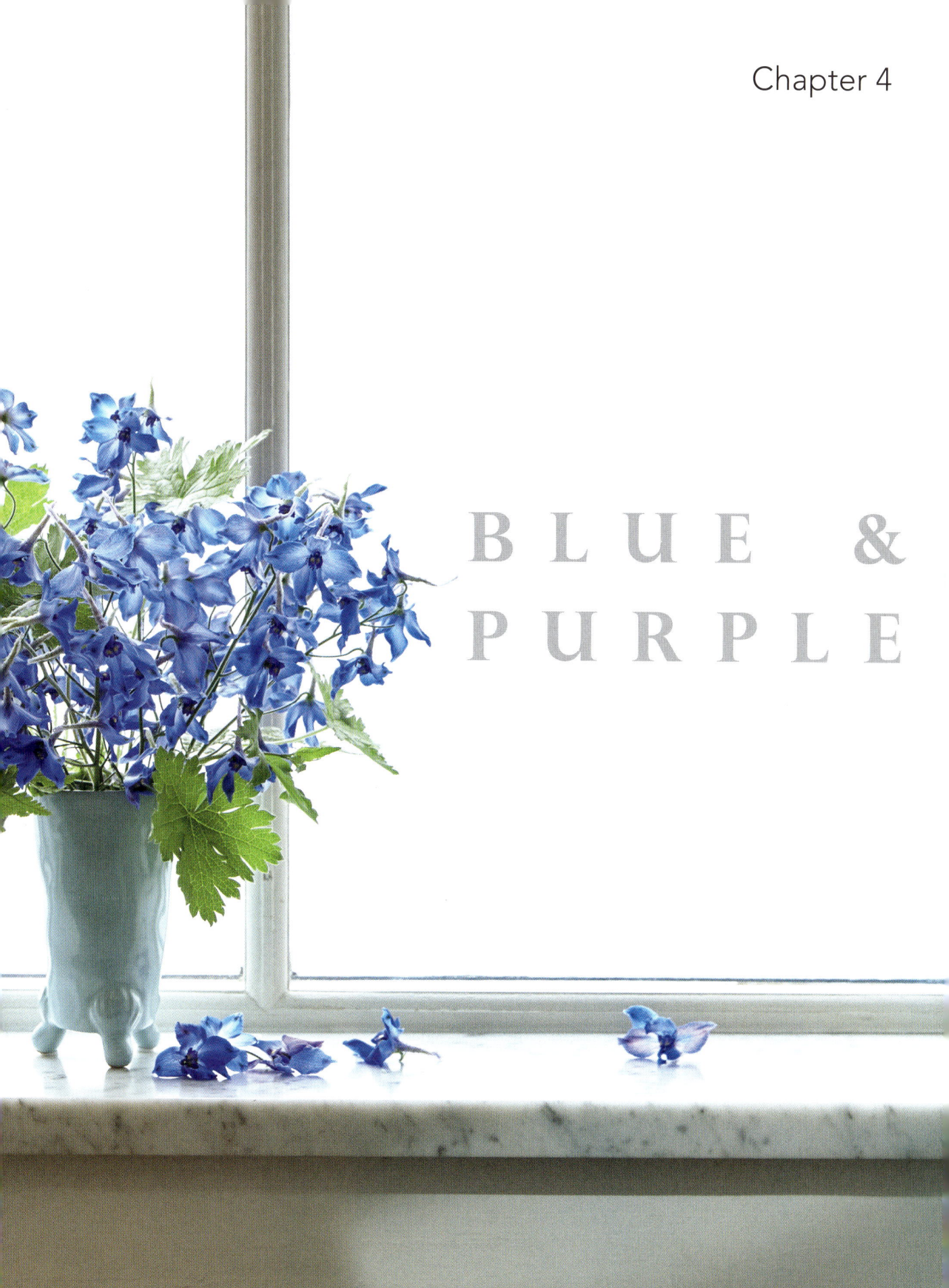

Chapter 4

BLUE &

PURPLE

FORGET ME NOT

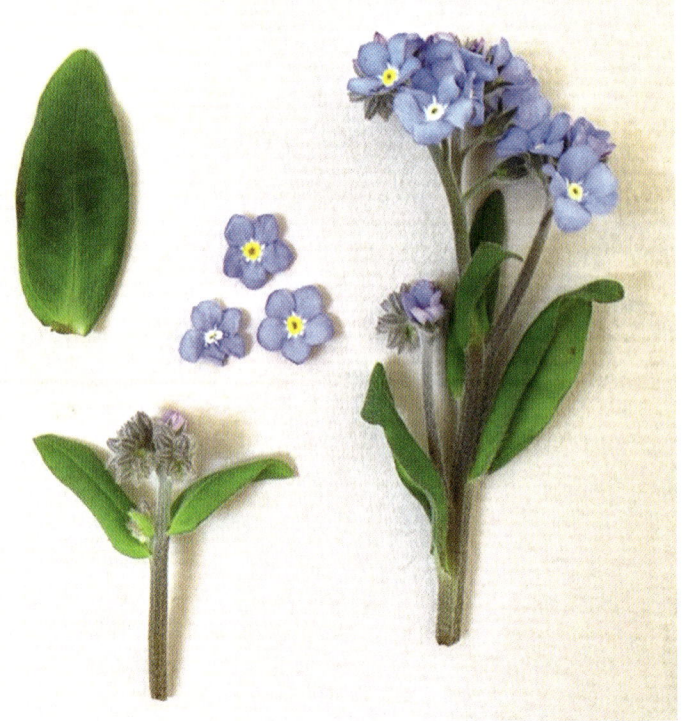

Latin name *Myosotis sylvatica*
Photograph shows Leaf, flowers, stem with leaves and flowers, stem with buds.
Collected Bushman Rock, Harare, Zimbabwe, 2015
Other common names Wood or garden forget-me-not
Family *Boraginaceae*
Origin Meadows and forest edges of Europe.
Description A herbaceous, spring to early-summer flowering biennial or perennial that grows to a height of 50cm.
Leaves These are alternate, narrowly obovate, 'hairy', and have with entire margins. The lowest leaves on the stem are stalked, while upper ones are not.
Flowers Bears cymes of small (up to 10mm across), five-lobed flowers. They can be blue, pink, or white, and have yellow and white centres.
Fruit Forms shiny and dark-brown fruits, up to 5mm long, which contain seeds.
Nomenclature *Myosotis* is from Greek *muosōtis*, meaning 'mouse's ear' referring to the furry leaves; *sylvatica* is from Latin and means 'pertaining to the forest'.
Uses The plant is an astringent and is used in herbal medicine to treat eye conditions. The leaves can be applied fresh or dried to wounds and can help staunch bleeding.

I picked these flowers from the garden my uncle's farm, Bushman Rock. The property nestles in the granite hills and balancing rocks just outside Harare and its vineyards slope down to the Nyamasanga River alongside a polo field established by my cousin after a stint in Argentina as a gaucho.

CHINESE FORGET ME NOT

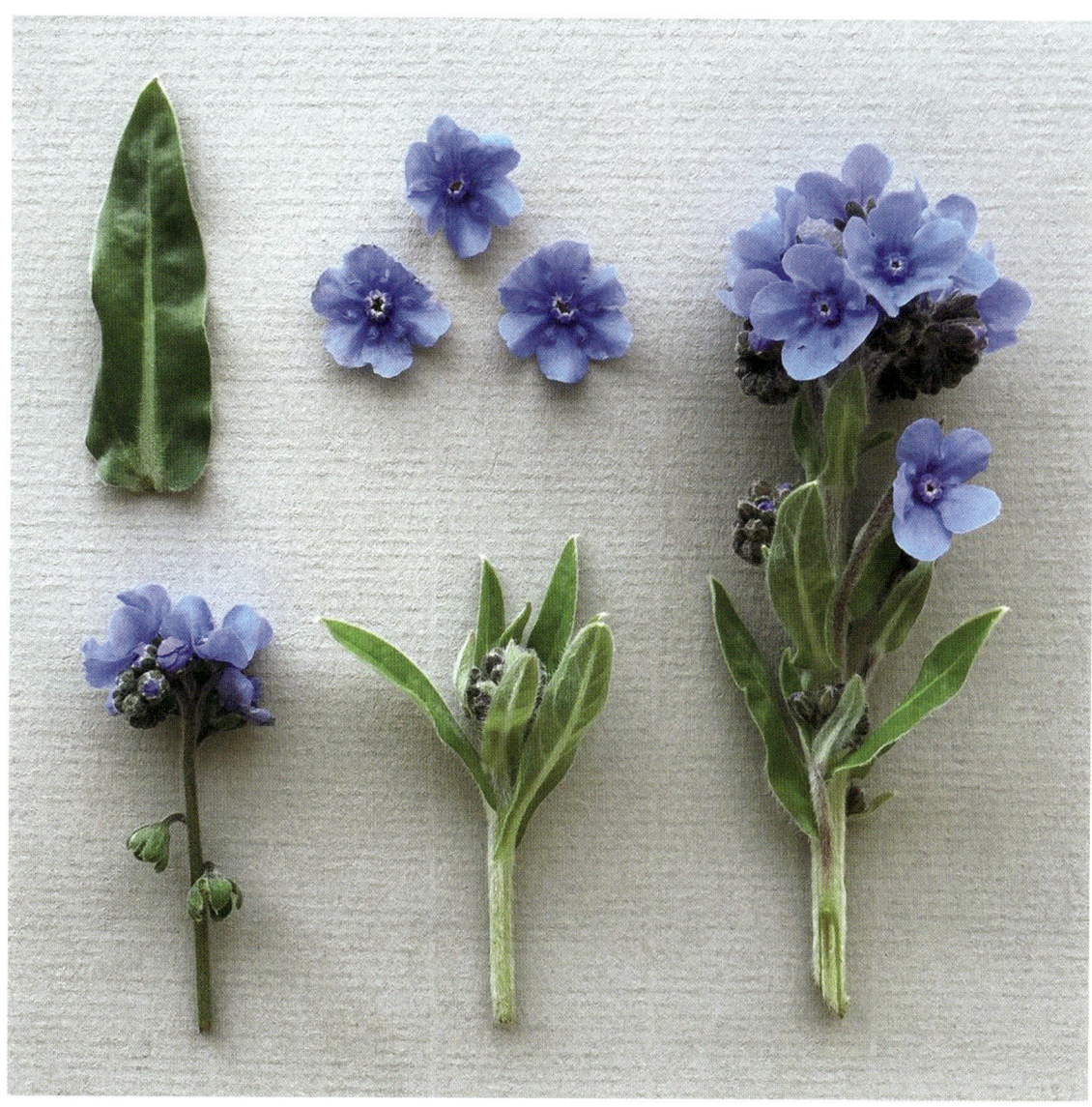

Latin name *Cynoglossum amabile*
Photograph shows Leaf, individual flowers, inflorescence, stem with buds, floret.
Collected Seoul, South Korea, 2015
Family *Boraginaceae*
Origin Southern Asia
Description A late-summer flowering, bushy annual, which grows to 45–60cm in height.
Leaves Lanceolate, hairy, grey-green leaves are up to 20cm long.
Flowers Bears inflorescences of showy, light-blue cymes of 'forget-me-not'-like flowers, that each measure 6mm in diameter.
Fruit Forms dry fruit that does not split open when ripe.
Nomenclature *Cynoglossum* is from a corruption of the Greek words *skylos* or *kynos* ('of the dog') and *glossa*, meaning 'tongue'. *Amabile* means 'loveable', or 'sweet and lovely'.
Uses The plant is used in traditional Chinese medicine to treat coughs as well as to stop the bleeding from wounds.

I bought these flowers from the flower market in Seoul's express bus terminal, Gyeongbuseon. Extraordinarily, the market is situated on the third floor of the bus station.

RHODES BUSH

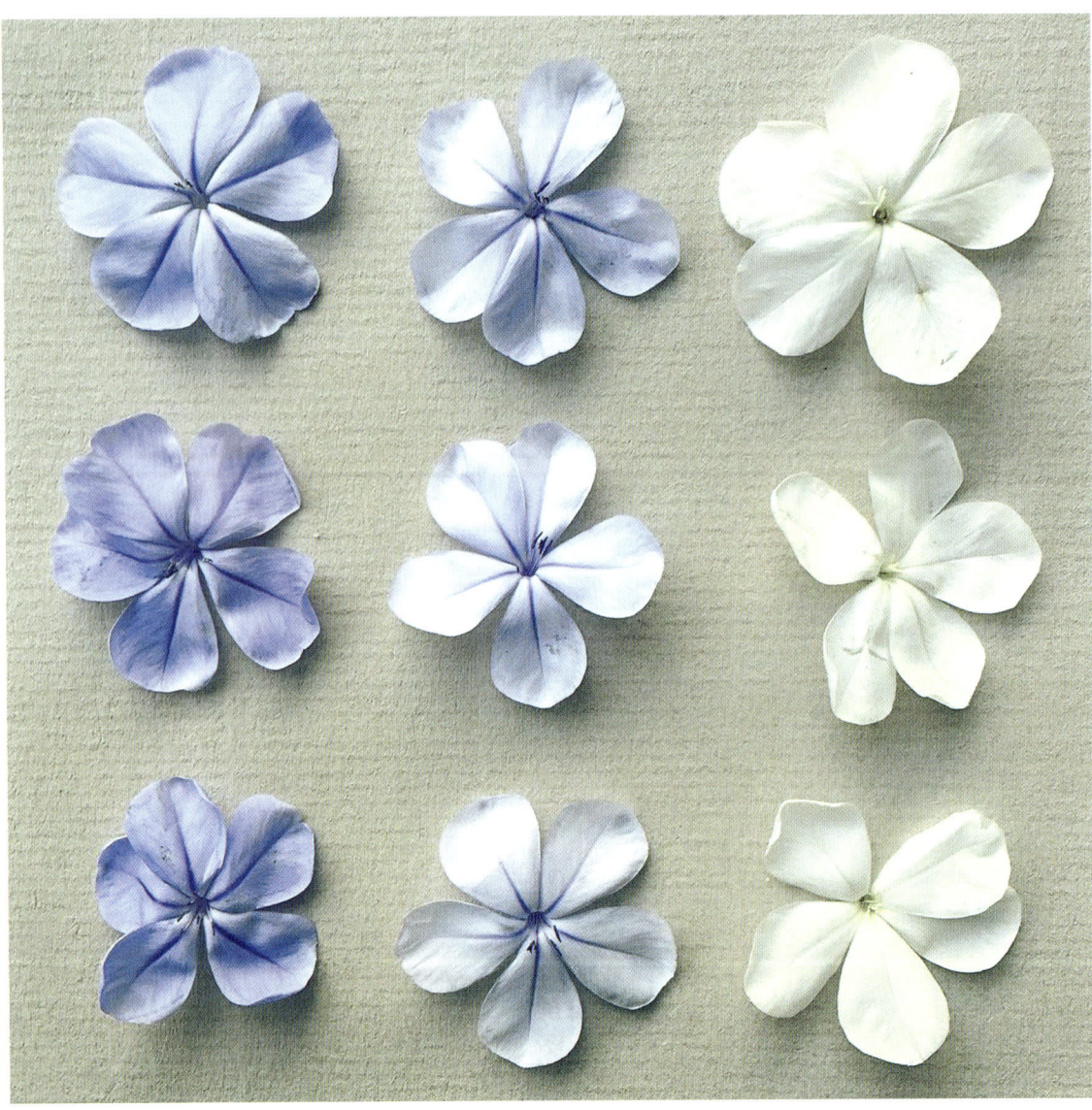

Latin name *Plumbago auriculata (syn. P. capensis)*
Photograph shows Flower heads
Collected Capetown, South Africa, 2015
Other common names Cape plumbago, Cape leadwort
Family *Plumbaginaceae*
Origin South Africa
Description Scrambling evergreen shrub that grows 1–3m tall with a spread of 2–3m. It flowers throughout the year in tropical climates, but only from summer through autumn in sub-tropical and warm, temperate climates.
Leaves Thin, light-green, spoon-shaped leaves are obovate with crenulate margins, and 10–25mm long. They have winged axils and ear-shaped appendages at the base.
Flowers Pale-blue to white, long-tubed flowers, each 15mm across, are borne in rounded, terminal clusters on dense racemes up to 15cm long.
Fruit Forms brown, elongated seed capsules, less than 5mm long.
Nomenclature Genus name comes from the Latin *plumbum*, meaning 'lead'. *Auriculata* means 'ear-shaped'. The epithet *capensis* for its original botanical name describes the Cape of Good Hope, South Africa, where the plant originates.
Uses Used in traditional medicine to treat wounds, broken bones, and headaches.

This was the favourite flower of Cecil John Rhodes (1853–1902), 'founder' of Rhodesia, now Zimbabwe. Rhodes planted them at his various African residences, hence it is commonly known as the 'Rhodes bush'

CORNFLOWER

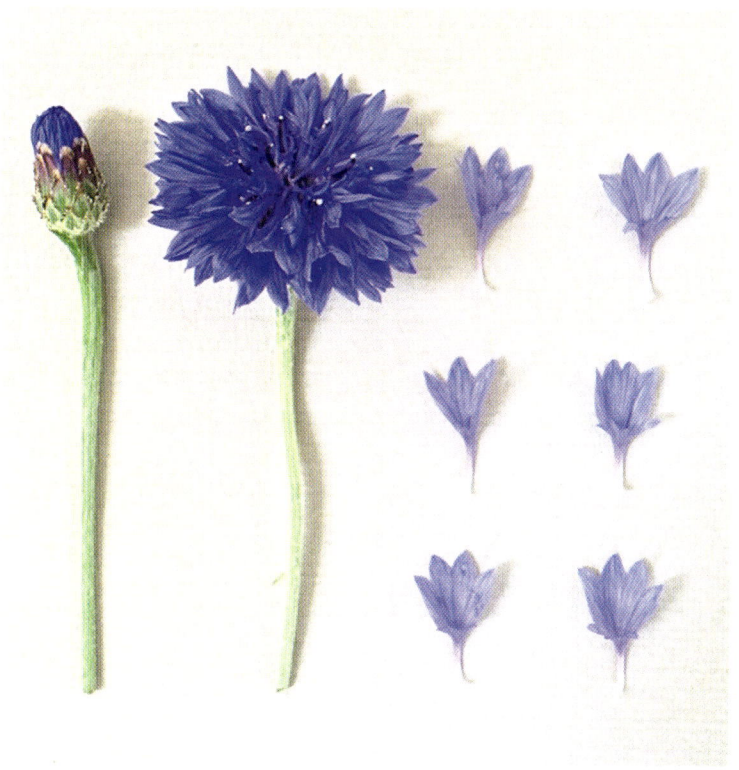

Latin name *Centaurea cyanus*
Photograph shows Bud, flower on a stem, ray florets.
Collected London, UK, 2014
Other common names Bachelor's button, Blue bottle
Family *Asteraceae*
Origin Europe
Description An upright annual that can be 40–90cm tall.
Leaves Forms mostly narrow, lanceolate leaves, 1–4cm long, that grow in whorls around the long stems; the lower leaves have curved lobes.
Flowers Bears intense blue flower heads, 3–4cm diameter, comprised of a ring of large, spreading ray florets surrounding a central cluster of disc florets.
Fruit Produces an elliptic, flattish, yellowish, fine-haired achene (3.5–4mm long), tipped with short bristles.
Nomenclature *Centaurea* is from *kentauros*, meaning 'centaur', from the Greek legend that the medical properties of a plant of this species were discovered by the wise centaur, Chiron. *Cyanus* means blue.
Uses Decorative ornamental. Dried petals are used in blends of tea including both Earl and Lady Grey.

The common name 'bachelor's buttons' refers to the folklore that cornflowers were worn as buttonholes by young men seeking love.

APOSTLE PLANT

Latin name *Neomarica gracilis*
Photograph shows Leaf, flower on a stem, sepals, single flower, style arms, ovary.
Collected Aberfoyle tea plantation, Zimbabwe, 2016
Other common name Walking iris
Family *Iridaceae*
Origin Central to South America – Mexico to Brazil.
Description Herbaceous perennial that forms large clumps from rhizomes and can grow to 90cm.
Leaves Sword-shaped (lanceolate), erect, ribbed, or heavily veined, leaves are 50–90cm long, and emerge from the soil in a clump. They are soft and flexible and bend gracefully backwards at the tip.

Flowers Bears scented, short-lived, iris-like, blue and white flowers in summer. The flower parts are in threes. The three petal-like, heart- to egg-shaped sepals, are white with brown bands near the base and the three smaller petals are bright blue with white streaks and tips that curl backwards.
Fruit None – plant spreads from the rhizomes.
Nomenclature Genus name is derived from the Greek *neo*, meaning 'new', and *Marica* the name of a Roman nymph. The specific epithet *gracilis* is Latin, meaning 'thin' or 'slender'. The common name comes from the belief that the plant will not flower until it has formed 12 leaves.
Uses Decorative ornamental.

Neomarica gracilis has naturalized in the forests surrounding the Aberfoyle tea plantation in Zimbabwe's Honde Valley. Its graceful flowers have an unexpected and lovely scent, not unlike Lily-of-the-valley.

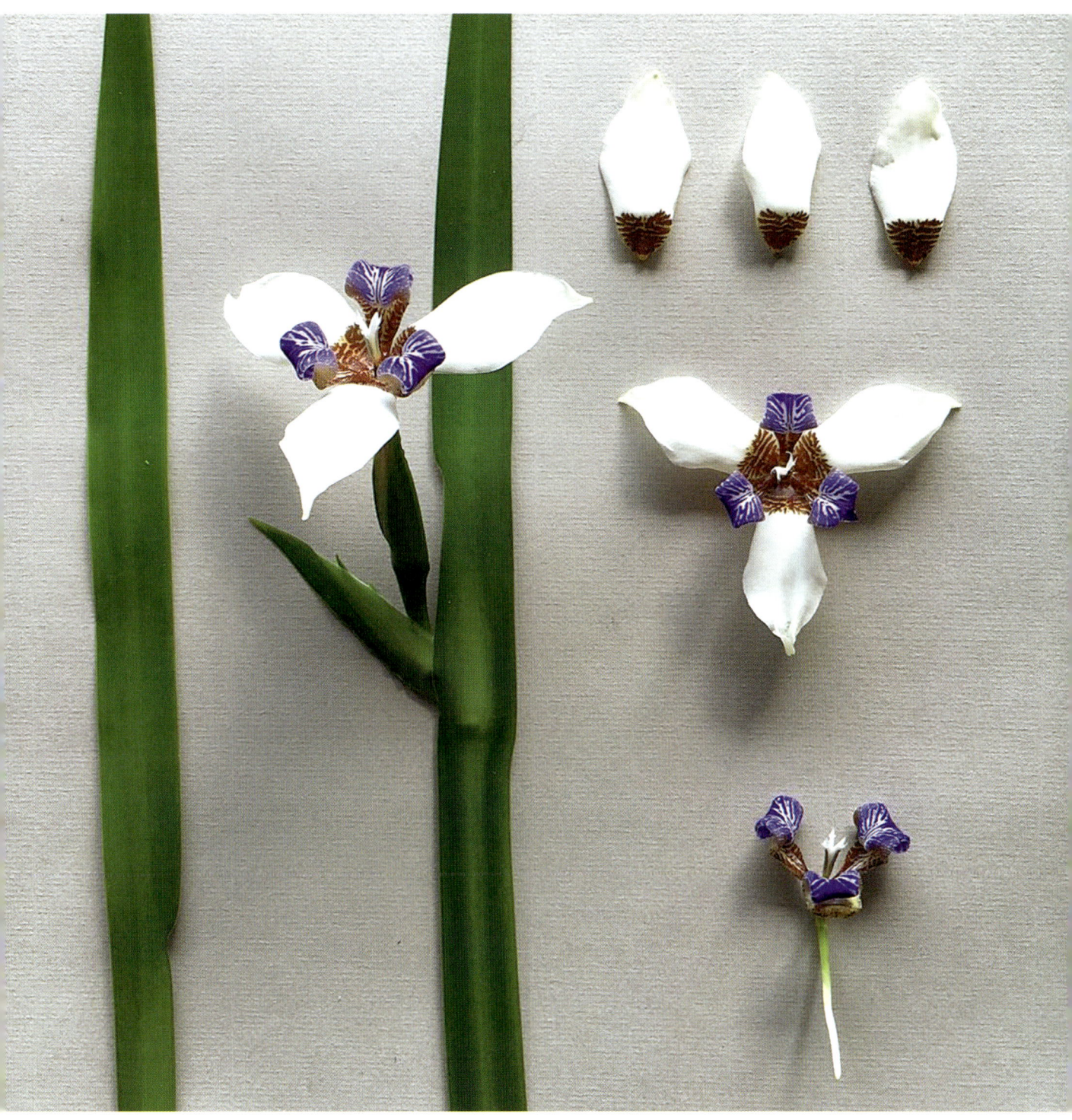

BEARDED IRIS

Latin name *Iris germanica*
Photographs show Right: fall petal, standard petal, styles, style arms. Opposite: flower on a stem.
Collected London, UK, 2015
Other common names German iris, Common iris, Liberty iris
Family *Iridaceae*
Origin Eastern Mediterranean
Description Herbaceous, spring-flowering perennial that grows up to 60–90cm from a rhizome.
Leaves Forms stalkless, usually slightly greyish-green, sword-shaped leaves with parallel veins and entire margins. The basal leaves are 30–70cm long, and stem leaves are smaller.
Flowers Bears inflorescences of two to four fragrant, dark-purple to pale-blue flowers (sometimes red, yellow, pink, or white, depending on the cultivar), 8–15cm wide. Flowers have six tepals, in two separate whorls – three outer tepals that curve downwards, and three inner broadly obovate, erect tepals of the same length. They also have three stamens, three styles, and large stigmatic lobes, shorter than inner tepals.
Fruit Forms elongated, three-lobed, capsules 4–5cm long, but rarely seed.
Nomenclature Genus is named after *Iris*, the Greek Rainbow Goddess; *Germanic* means German.
Uses The roots of this species along with *I.pallida* and *I. florentina*, are used in perfumery for their delicate 'powdery' violet scent and fixative properties.

Iris scent is a very time-consuming material to produce. The rhizomes, known as orris roots, are collected, peeled, and aged for up to four years in a curing process that develops the desired properties – the longer the roots are stored, the higher the concentration of the scent. Once the level of ageing has been achieved, the rhizomes are pulped. The pulp is then put through a process of steam distillation to extract an essential oil. As the 'oil' cools it hardens to form a waxy substance, or 'concrete', often referred to as iris butter.

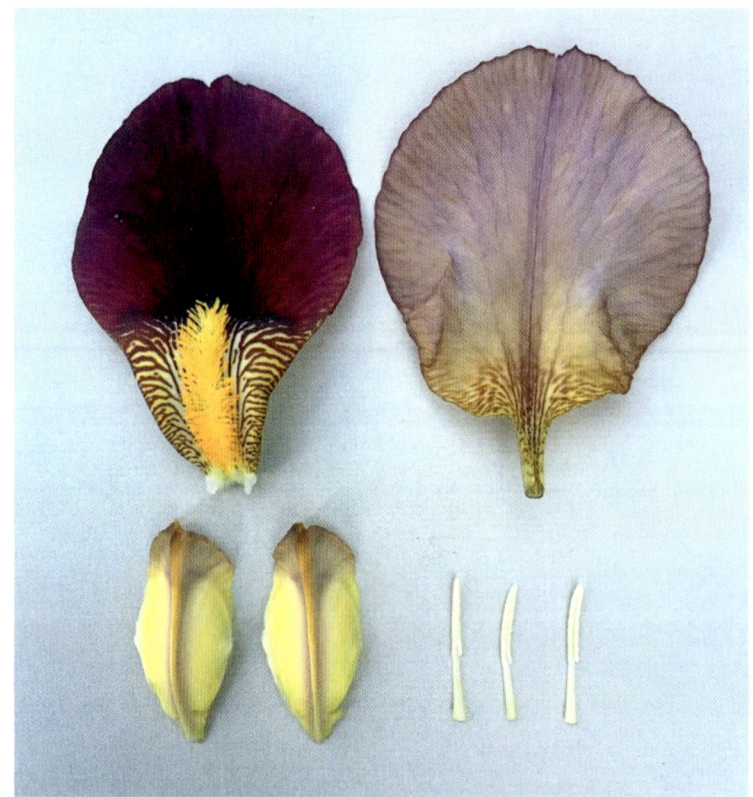

As a boy, I would collect and dry various botanicals from the garden. I often mixed dried rose petals, geranium, and lemon verbena leaves together, then 'fixed' the resulting *pot pourri* with a sprinkling of white orris root powder from a bag that has been sitting on a shelf in the pantry for as long as I can remember.

The little plastic bag of cream-coloured orris powder sat on a shelf next to a canister of cream of tartar (which we were led to believe came from the seeds of Baobab trees), and a medicinal-looking brown bottle that contained acetic acid – a drop of this was added to the Christmas cake icing each year to prevent it from cracking. Whenever I was in the pantry I always opened the brown bottle and took a whiff of the acid's acrid scent, which felt like it went right up my nose.

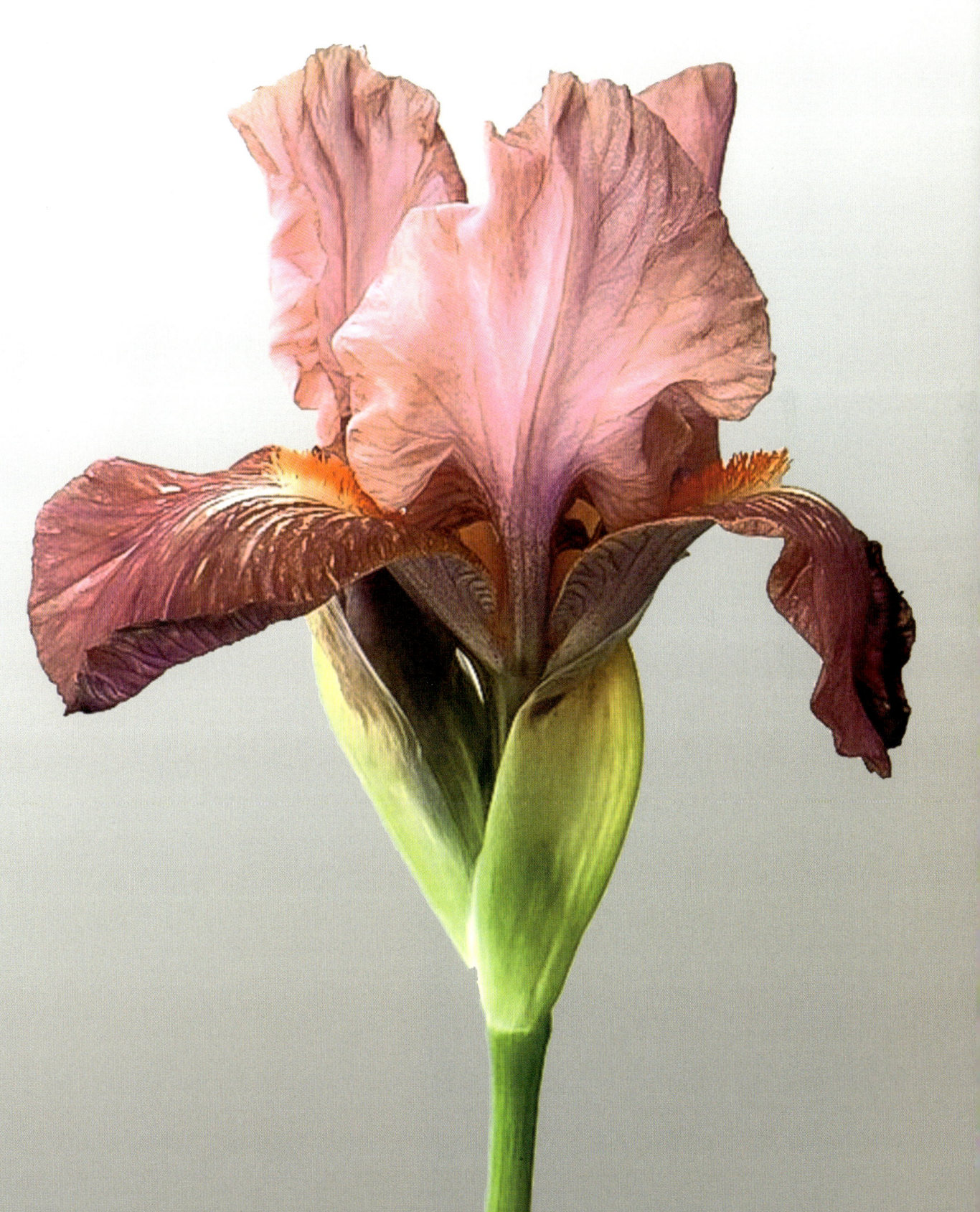

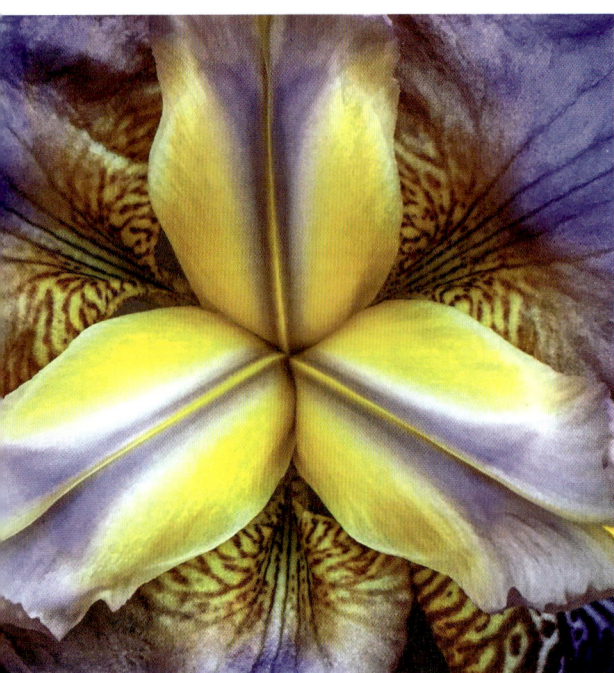

Looking directly into the centre of the flower, the style arms are clearly visible.

The Bearded iris was first given its scientific name *Iris germanica* by the 18th century Swedish botanist Carl Linnaeus (1707–78). Since then thousands of cultivars have been developed in many colours. They are divided into six groups: miniature, dwarf bearded; standard dwarf bearded; intermediate bearded; border bearded; miniature tall bearded; and tall bearded. I collected the majestic tall-bearded specimens shown here in 2015, in London, UK.

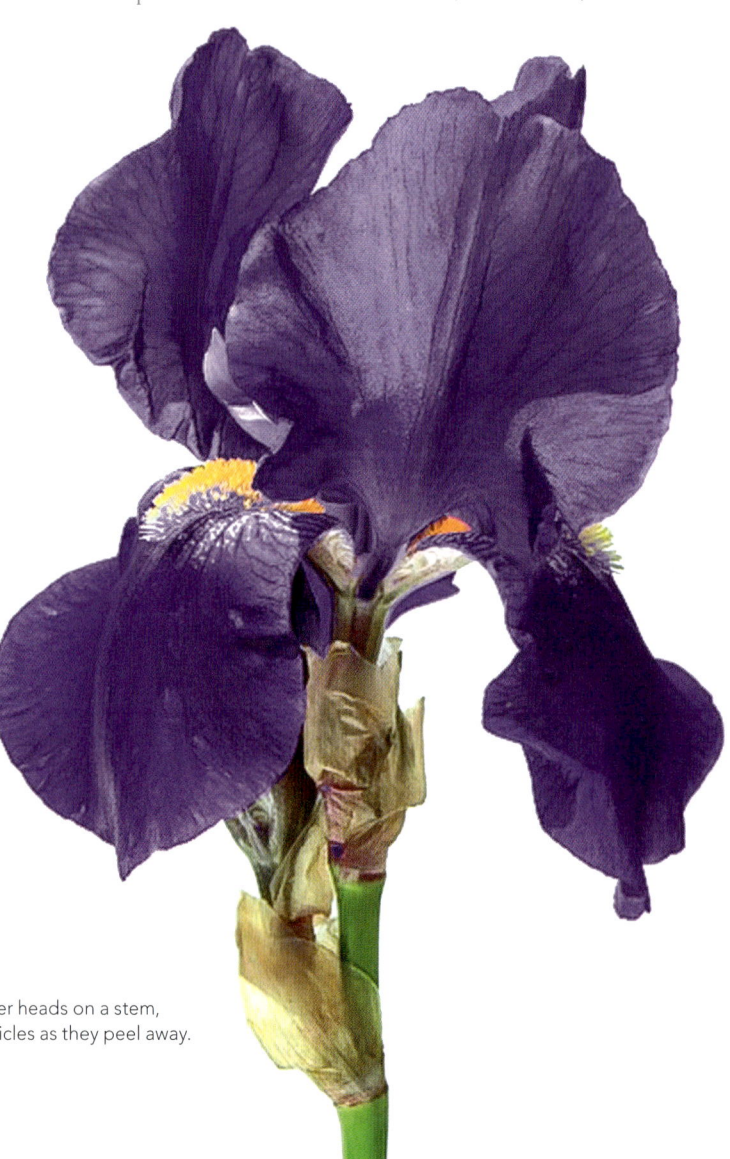

Spectacular flower heads on a stem, showing the pedicles as they peel away.

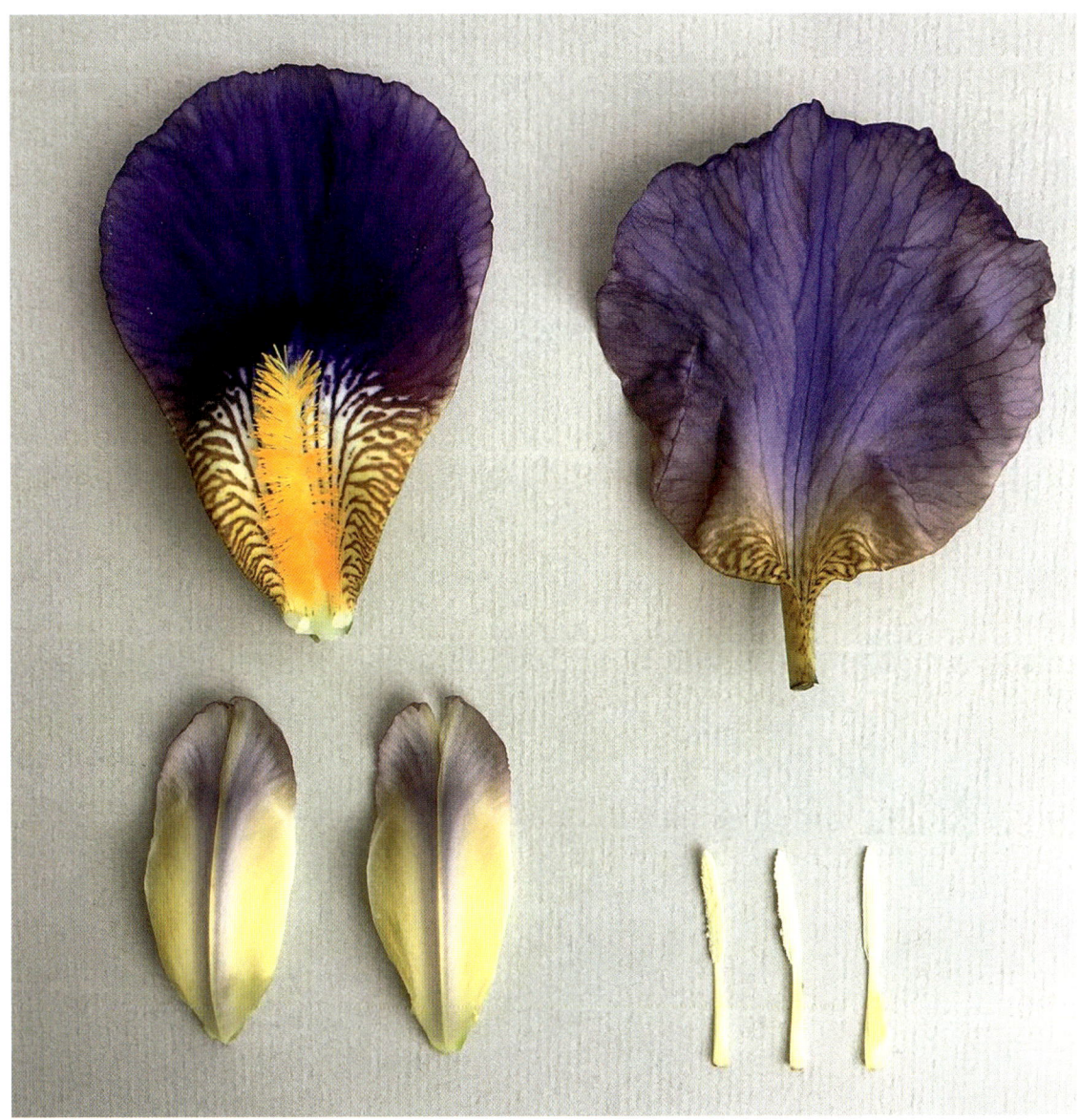

Dissection of an *Iris germanica*, showing
clockwise from top left: fall petal,
standard petal, three stamens, and two
style arms.

MANCHURIAN VIOLET

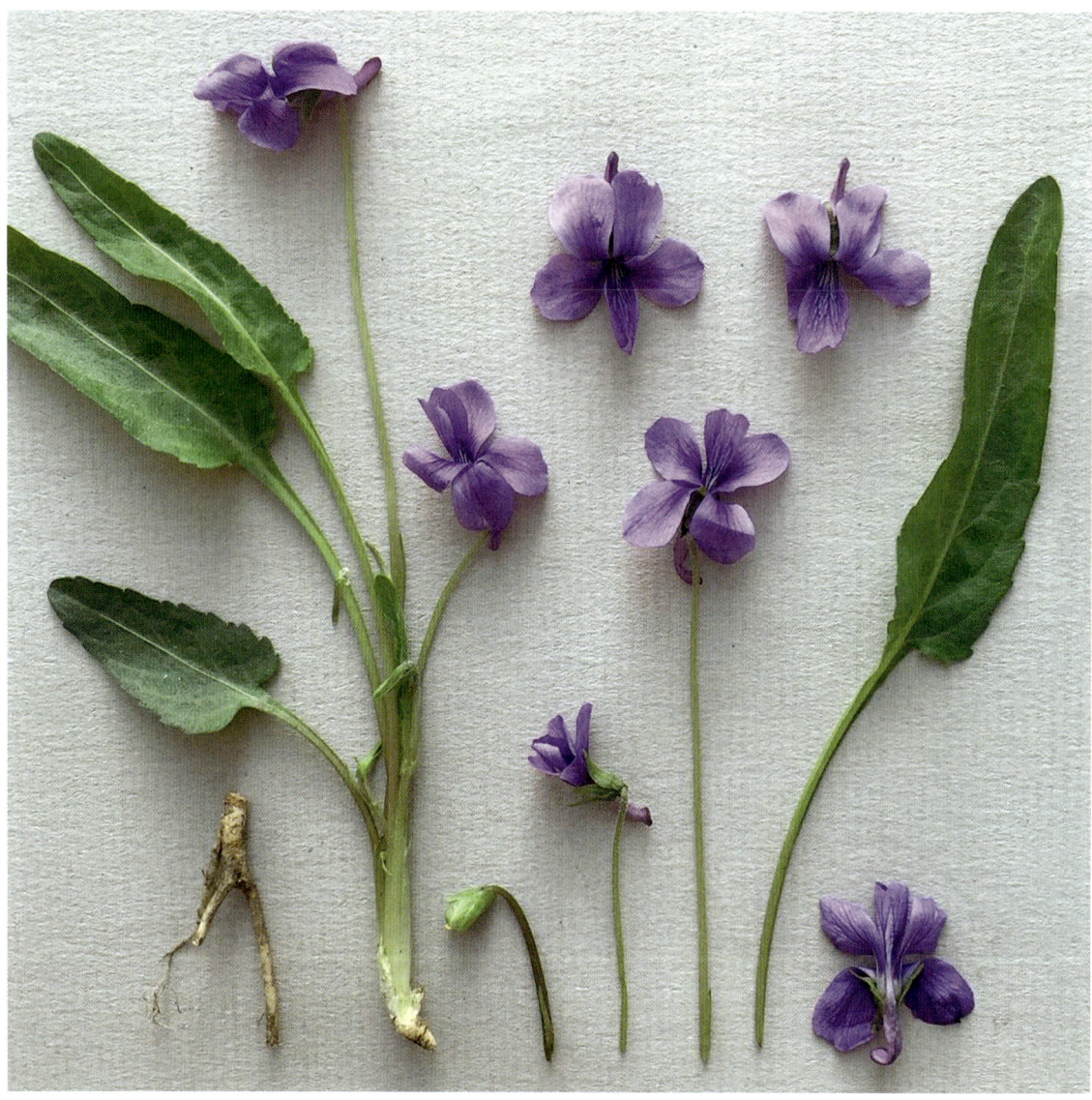

Latin name *Viola mandshurica*
Collected Seoul, South Korea, 2016
Photograph shows Section of root, plant, flower heads, leaf, flower on stem, buds.
Other common names Northeastern violet (China), Jebikot (Korea)
Family *Violaceae*
Origin Eastern Asia – Japan, China, Korea.
Description Stemless, flowering perennial that grows up to 20cm in height. It blooms late winter to early spring, and leaves and flowers emerge directly from a rhizome.
Leaves Oval to lance-shaped leaves with winged petioles are 2–7cm long.
Flowers Trumpet-shaped, purple,

bisexual blooms have five petals and are 2.5–3cm long.
Fruit Forms oblong capsules 1–1.5cm long, which contain numerous red-brown seeds.
Nomenclature The genus name *viola* originates from Vion a beautiful mythological Greek priestess. The specific epithet *mandshurica*, means 'from Manchuria', an area of the species' native habitat.
Uses In Korea, the blooms are used to colour special 'flower' pancakes for their Samjinnal festival, a celebration of spring. Recent research has also shown that preparations of the plant may have potential in the treatment for asthma.

Taking a walk in downtown Seoul I noticed a little clump of purple flowers growing out of a crack in the pavement. Although the flowers were instantly recognizable as violets, the lance-shaped leaves were a surprise as I am more familiar with the round leaves of the sweet violet, *Viola odorata*.

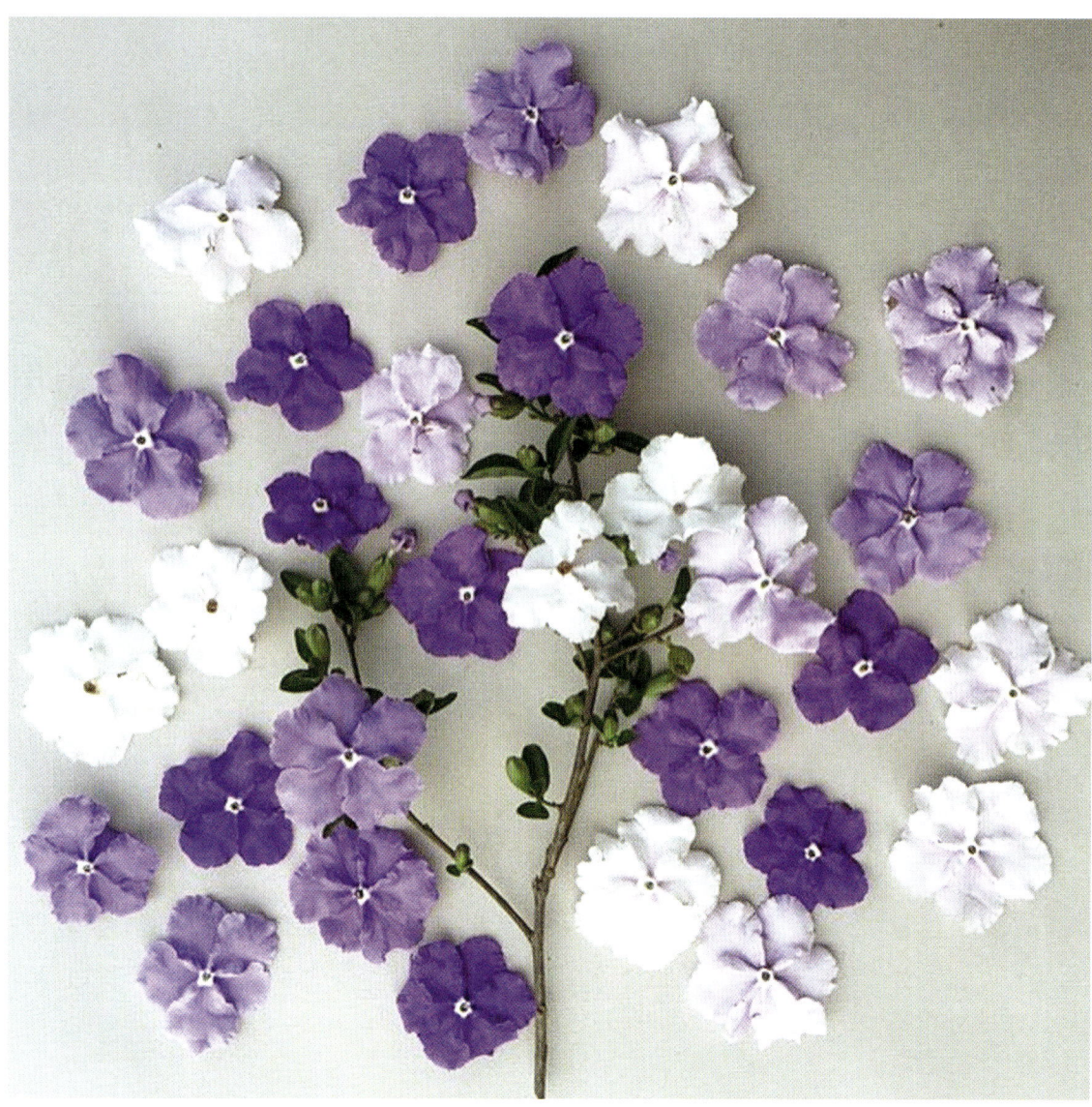

Latin name *Brunfelsia grandiflora*
Photograph shows Flowers at different colour stages, stem with young leaves.
Collected Aberfoyle, Zimbabwe, 2015
Other common names Royal purple brunsfelsia, Kiss-me-quick
Family *Solanaceae*
Origin Tropical Central and South America – Brazil, Bolivia, Peru, Ecuador, Venezuela, Colombia.
Description Evergreen, perennial shrub, usually 1.5m tall with a spread of 1m, but can grow to 3m with a spread of 2m.
Leaves Simple, alternate, dark-green, oblong leaves with entire margins are up to 20cm long.
Flowers Short-lived, trumpet-like, fragrant flowers have five petals and can be up to 10cm long. They are initially violet, before fading to lavender, and finally white over three to four days.
Fruit Bears yellow fruits that are poisonous to humans, but edible to birds.
Nomenclature The botanist Carl Linnaeus named the genus after German botanist Otto Brunfels (1488–1534). The epithet *Grandiflora*, meaning 'large flower', refers to the size of the blooms.
Uses Although the plant is poisonous to humans, extracts from the root can be used in the treatment of rheumatism and for some snake bites. Amazonian tribes are known to add leaves, roots, and bark to the ayahuasca-bark hallucinogenic brews prepared for their religious ceremonies.

Wonderfully present in my Zimbabwean childhood was the sweetly scented perfume of the Yesterday-today-and-tomorrow plant. As the individual flowers last only a few days, and are initially purple, then lilac, then near-white, the bushes are always an array of blooms in all three colours, hence one its common names.

SNAKE'S HEAD FRITILLARY

Latin name *Fritillaria meleagris*
Photograph shows Leaf, flower on a stem, petals, stamens, pistil, bud.
Collected London, UK, 2016
Other common names Guinea hen flower, Chequered daffodil
Family *Liliaceae*
Origin Temperate regions of Europe and western Asia.
Description A bulbous perennial that grows to 30cm in height.
Leaves The glaucous, lance-shaped, grass-like, grey-green leaves can be 6–13cm long.
Flowers Bears bell-shaped, pendant flowers with chequered petals; colour varies from reddish-brown to purple or white, depending on cultivar.
Fruit Forms erect, flat-topped, six-sided capsules.
Nomenclature Genus name *Fritillary* comes from the Latin, *fritillus*, meaning 'dice-cup', referring to the markings on the petals of some species. *Meleagris* is Latin from Greek and means 'guinea fowl' referring to this plant's spotted petals, like the markings of the bird.
Uses Decorative ornamental. The plant had a reputation as a healing herb, but is no longer used medically.

I find these meadow flowers enchanting. I entirely agree with British author and gardener Vita Sackville-West who said of them 'The fritillary looks like something exceedingly choice and delicate and expensive, which ought to spring from a pan in a hothouse, rather than share the fresh grass with buttercups and cowslips.'

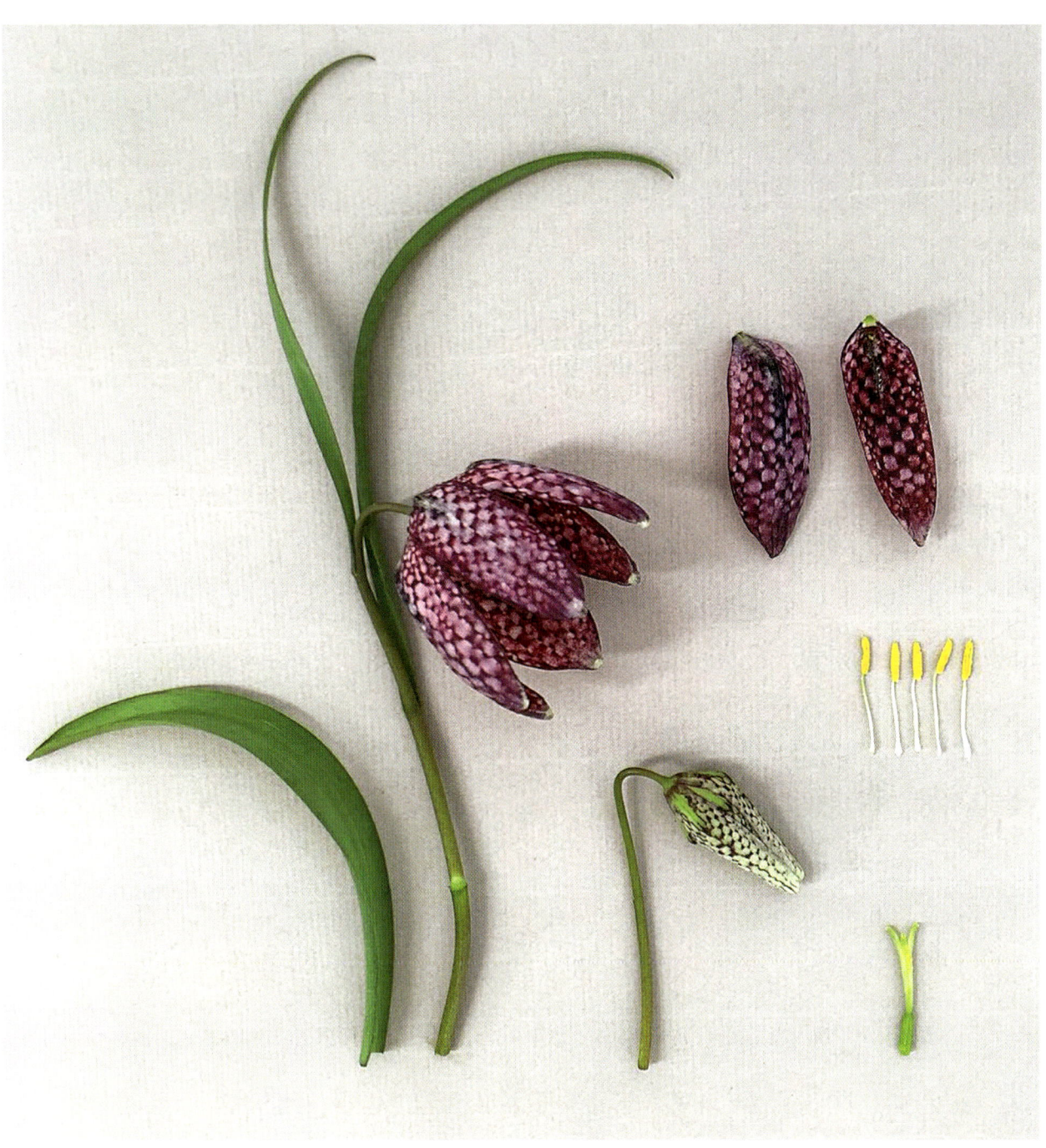

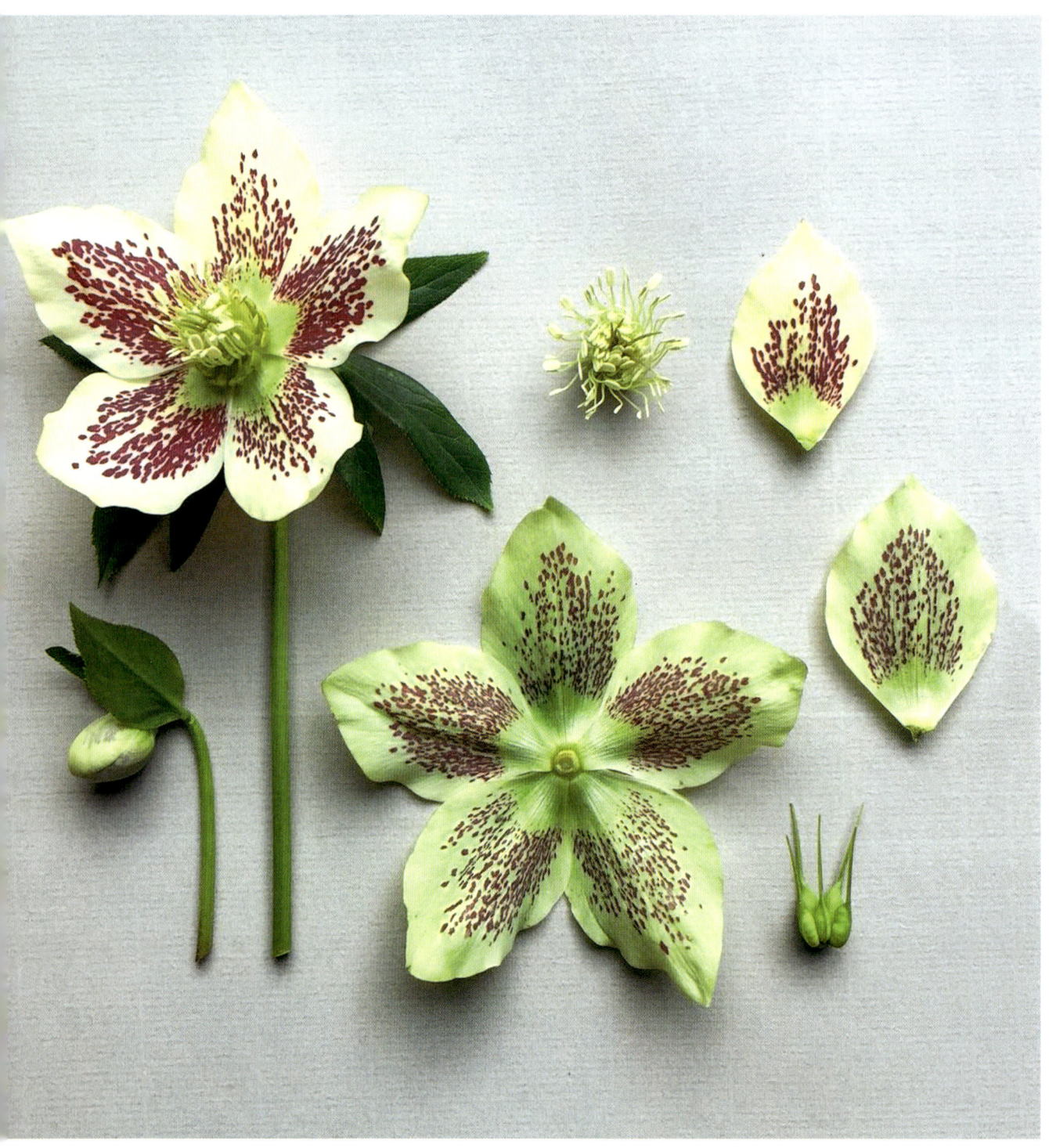

LENTEN ROSE

Latin name *Helleborus orientalis*
Photographs show Opposite: bud, flower, corona (anthers and pistils), petals, fruit underside of flower. Right: bud and leaves, stem with flower and bud, corona, anthers, petals. Below right: flowers, corona, petals, fruit, underside of flower.
Collected London, UK, 2016
Family *Ranunculaceae*
Origin Scrub woodland and rocky sites in southern and central Europe, from Slovenia to Macedonia.
Description Erect, evergreen perennial that grows 30–90cm tall from rhizomes.
Leaves Leathery, shiny, dark-green leaves have 'toothed' margins.
Flowers The saucer-shaped, lavender, pink, and white solitary winter flowers are about 6cm across.
Fruit The fruits are follicles.
Nomenclature The genus name *Helleborus* means 'food for a fawn', from the Greek words *ellos/hellos*, meaning 'fawn' and *bora* meaning 'food'. Alternatively, *hele* means 'to take away' so it could also be described as 'take way food', which refers to the emetic effect of the plant. The epithet *orientalis*, means 'from the East'.
Uses Decorative ornamental. *Helleborus orientalis* sub sp. *orientalis* is used for weight loss in Russian medicine.

I collected these specimens from a garden I often frequent in Onslow Square, South Kensington. Blooming in late winter, through to early spring, the Lenten rose bears masses of multi-hued speckled flowers, and is a rare joy when there is not much colour in the garden.

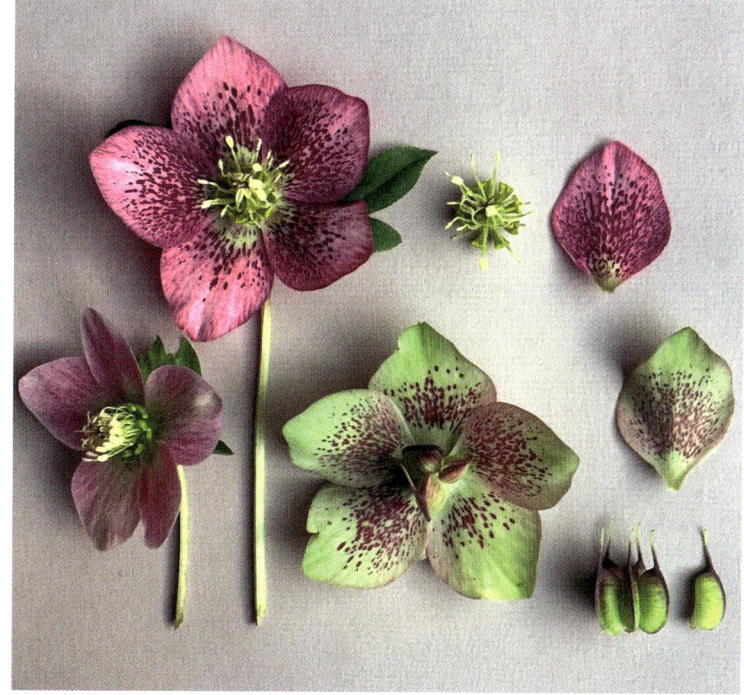

PASSION FLOWER CLEMATIS

Latin name *Clematis florida* 'Sieboldii'
Photograph shows Flower, central boss (with stamen), leaves, tepals with stamen
Collected Virginia, USA, 2016
Family *Ranunculaceae*
Origin South and southeast China.
Description A deciduous to semi-evergreen climber that is adapted to the cold, and grows up to 2.5m in height.
Leaves Dark-green, lanceolate to ovate leaves on twinning stalks are made up of three to nine leaflets. Leaves turn purplish in autumn.
Flowers From mid-summer to early autumn, plant bears flowers resembling those of a passion fruit with creamy tepals and a frilly, purple central boss, or stamen, that can be up to 10cm in diameter.
Fruit Forms fluffy seed heads.
Nomenclature *Clematis* is from Ancient Greek *klimatitis* meaning 'climbing plant'. Cultivars are often named after their originators or characteristics. This variety is named after the Dutch zoologist Phillipp van Siebold (1796–1866) who brought it to Europe from Japan.
Uses Decorative ornamental.

I collected this specimen from the garden of American friends Evan and Holly Chapple. Holly is a force of nature and visionary in the floristry world. Trips to Hope, their flower farm in Virginia, to 'Flower stock' (think all-day flower arranging, live music, and ho-downs) are an autumn highlight for me.

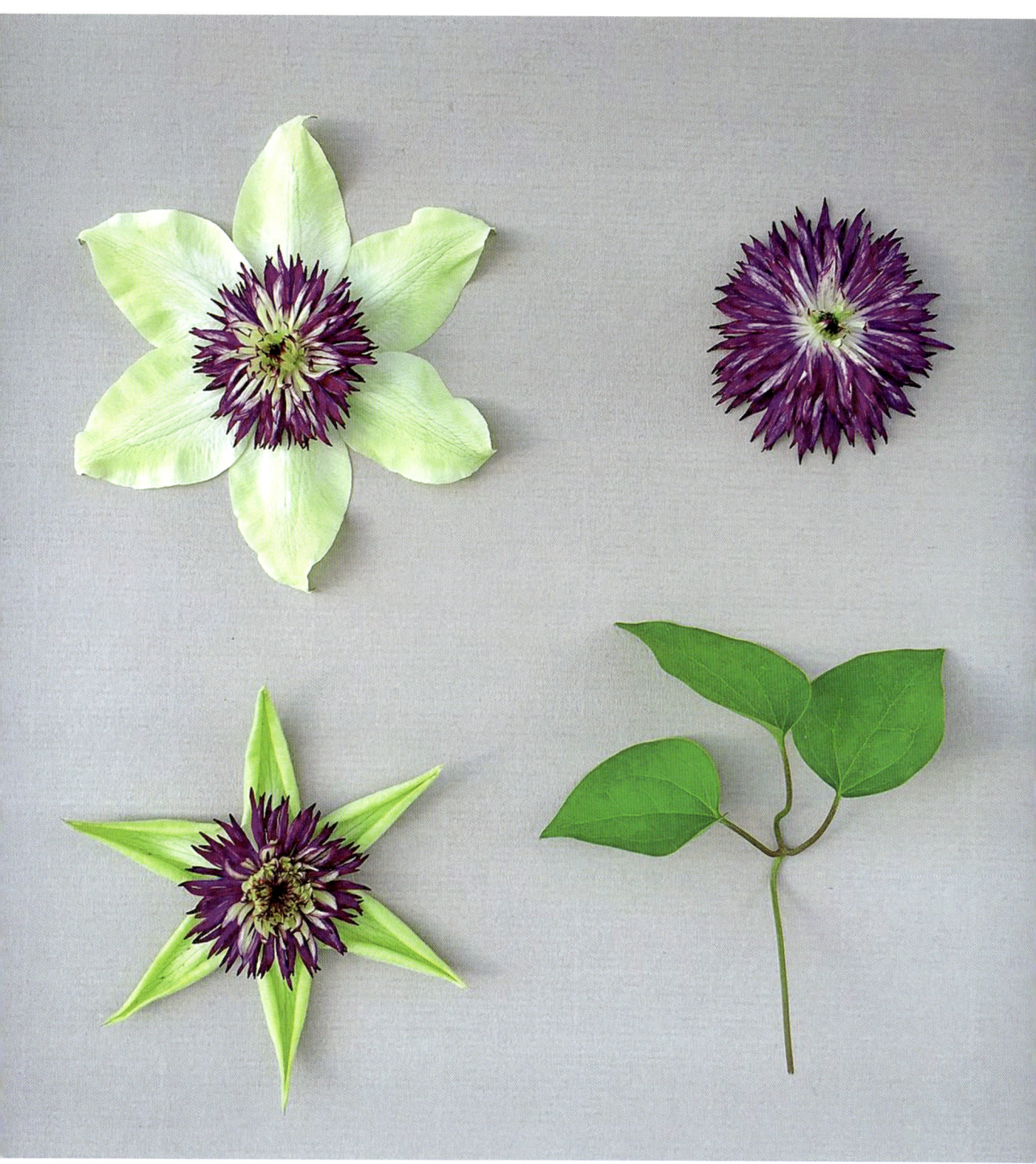

LAVENDER MIST VANDA

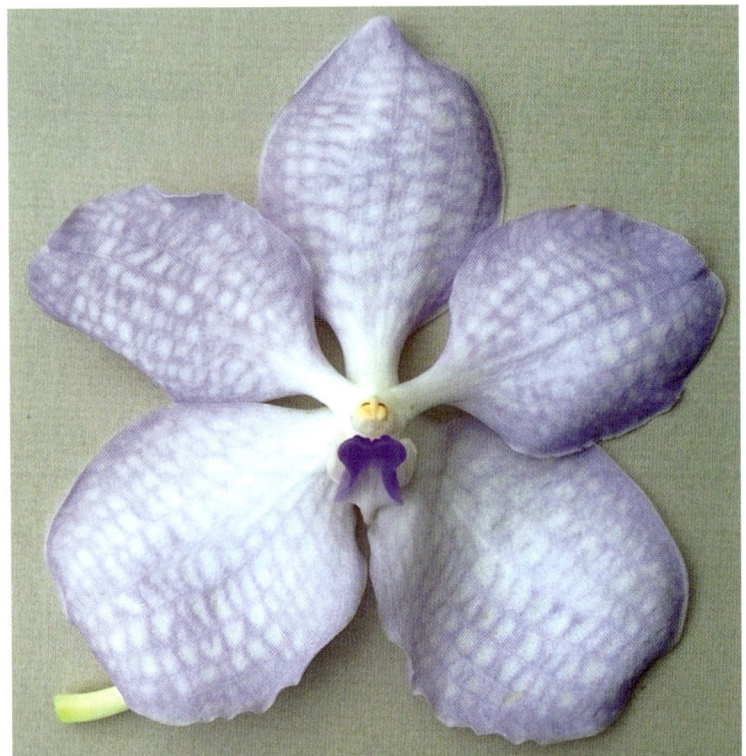

Latin name *Vanda kanchana* 'Lavender mist'
Photographs show Left: flower head. Below left: column, sepals, petals, anther cap, pollinia.
Collected London, UK, 2014
Other common name Vanda orchid
Family *Orchidaceae*
Origin A hybrid, of which one parent plant, *Vanda coerulea*, is native to China.
Description Epiphytic plant of the orchid family that blooms frequently.
Leaves Leathery, strap-like leaves are attached to main stem by leaf sheaths.
Flowers Borne on a terminal spike, the flower's light-purple petals are tessellated, with a distinctive, dark-purple lip.
Fruit None
Nomenclature The name *Vanda* comes from the Sanskrit name for this species. 'Lavender Mist' is the cultivar name of this hybrid.
Uses Decorative ornamental.

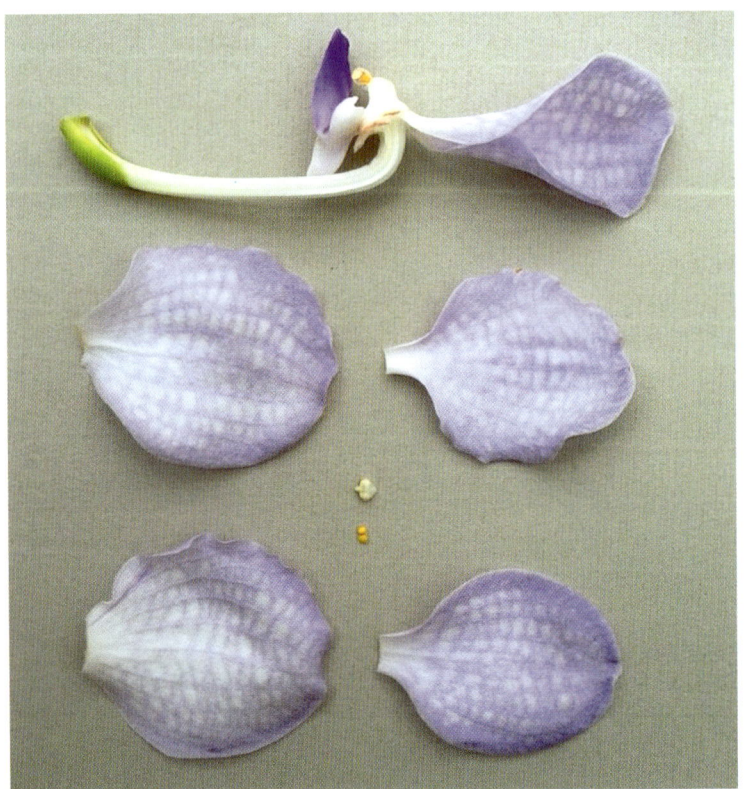

When, at the age of 11 years, I joined the Orchid Society of Zimbabwe, I was their youngest paid up member. My first orchid was an epiphytic *Vanda coerulea* that cost me a month's pocket money. This same plant still thrives, growing on the trunk of one of the Natal mahogany trees (see page 60) at Maduma where I first positioned it 30 years ago.

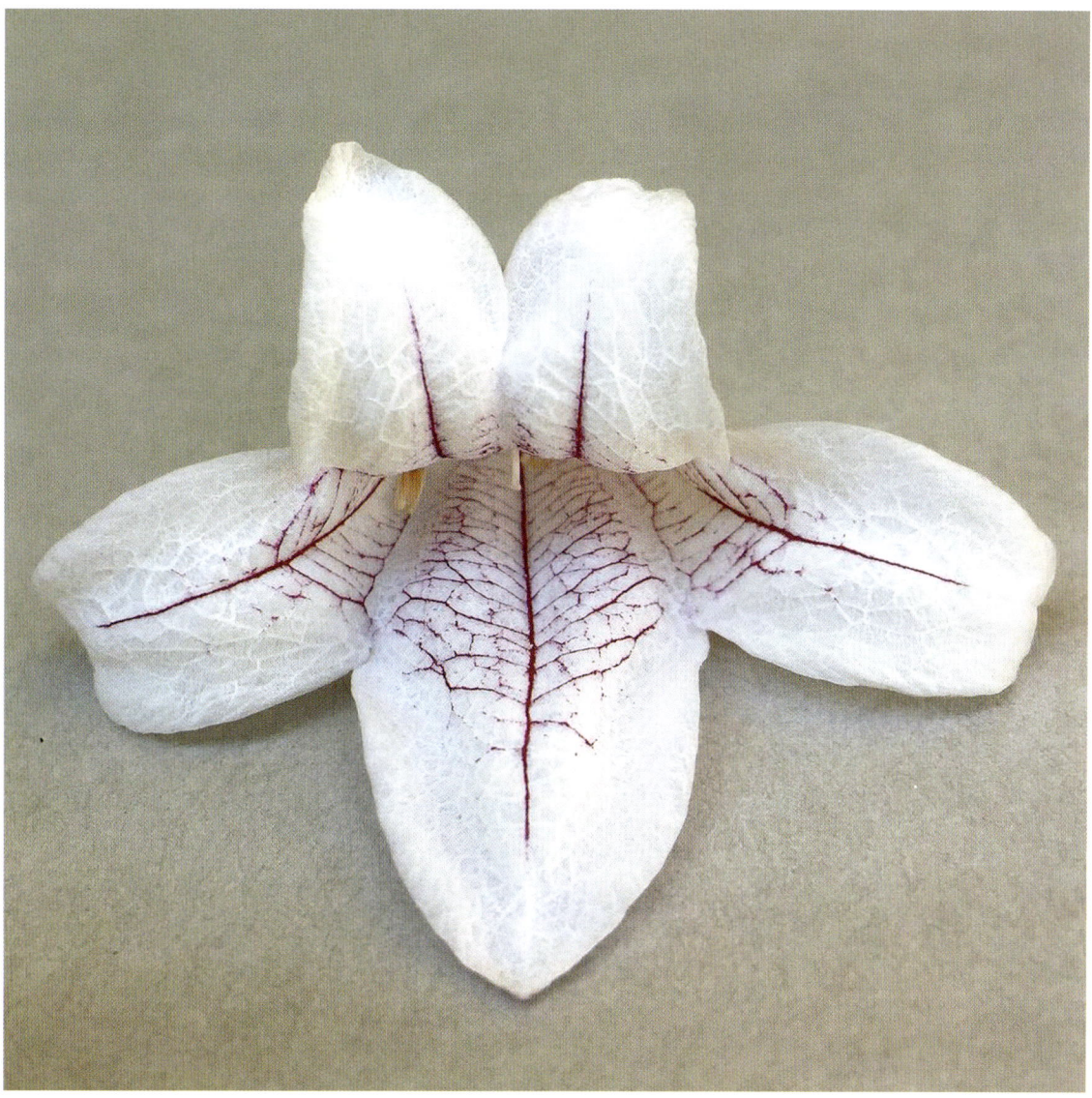

Latin name *Mackaya bella*
Photograph shows Flower
Collected Maduma, Zimbabwe, 2015
Other common name Forest bell bush
Family *Acanthaceae*
Origin Subtropical to temperate regions of southern Africa – Eastern Cape, Kwazulu Natal, Northern Province, Swaziland.
Description An evergreen, semi-hardy shrub that grows up to 4m tall.
Leaves Simple, oppositely arranged dark-green, glossy leaves have prominent veins and midribs, and wavy, toothed margins. Small, hairy pockets are often found in the axil of the veins.
Flowers The large, mauve flowers are borne on terminal racemes. Blooms are marked with disctinctive purple lines (veins) and can be up to 5cm across.
Fruit Forms green, club-shaped capsules that mature to brown, then explode into two sections to eject the seeds.
Nomenclature The genus *Mackaya* is named after Scottish-born botanist and author of *Flora Hibernica*, James Townsend Mackay (1775–1862). *Bella* is Latin, meaning 'beautiful'.
Uses Decorative ornamental. The timber can be used to light fire by friction.

Included in the indigenous garden I planted at Maduma, the flowers of this shrub are irresistible to butterflies and bees.

CHINESE WISTERIA

Latin name *Wisteria sinensis*
Photograph shows Raceme, leaf, open flowers, buds.
Collected Maduma, Harare, Zimbabwe, 2016
Other common name Wisteria
Family *Fabaceae*
Origin China
Description A vigorous, deciduous, spring-flowering, ornamental woody climber that can grow up to 7m with support.
Leaves Compound, alternate leaves, taper strongly at the tips and appear after flowers. They are 9–11cm long and made up of seven to 13 leaflets.
Flowers Bears pendulous racemes (up to 50cm long), of fragrant, lilac and purple, pea-like flowers in spring.
Fruit Forms 15-cm-long seed pods that contain up to three flat, round, seeds, which are poisonous to humans.
Nomenclature British botanist Thomas Nuttall (1786–1859) said the plant was named in honour of American anatomist Dr Caspar Wistar (1761–1818); but it is also thought he may have named it after his friend Charles Jones Wister. *Sinensis* means Chinese or China.
Uses Decorative ornamental. In traditional Chinese medicine the seed of *Wisteria sinensis* is used as a diuretic. The Chinese also make a local delicacy called 'Teng Lo' by preserving the flowers in sugar and flour. Fibres taken from the stems can be used to make paper.

I am often asked what my favourite flower is. I generally respond by saying it depends what month it is. However, truly, it is those of *Wisteria sinensis* and this goes right back to my childhood in Zimbabwe.

In the garden of Maduma, the house where I grew up, grows a spectacular gnarled old *Wisteria sinensis* planted many years ago by my grandmother. Unlike in the UK where almost every street has at least one wisteria growing up a house, they are a rarity in Zimbabwe, and the brevity of the blooming makes the sight all the more special. In the brief African spring, the bare winter branches of the wisteria spike with an abundance of inflorescences, which, coaxed by the warmer weather, elongate and open into cascades of scented purple flowers. After school in spring I could often be found perched on top of the courtyard wall over which the wisteria sprawled, surrounded by a sea of lilac-coloured flowers, 'drunk' on their scent, and the utter beauty of it all.

The opulence of wisteria on the streets of London allows me to relive those moments each spring. There are some particularly spectacular specimens that I make sure I try to catch in bloom each year; a few of them have become old friends.

I continue to be captivated by this plant, and a while back I initiated *#wisteriawatch* on social media to track the blooming of wisteria around the globe, as specimens in the southern and northern hemispheres bloom at opposite times of the year.

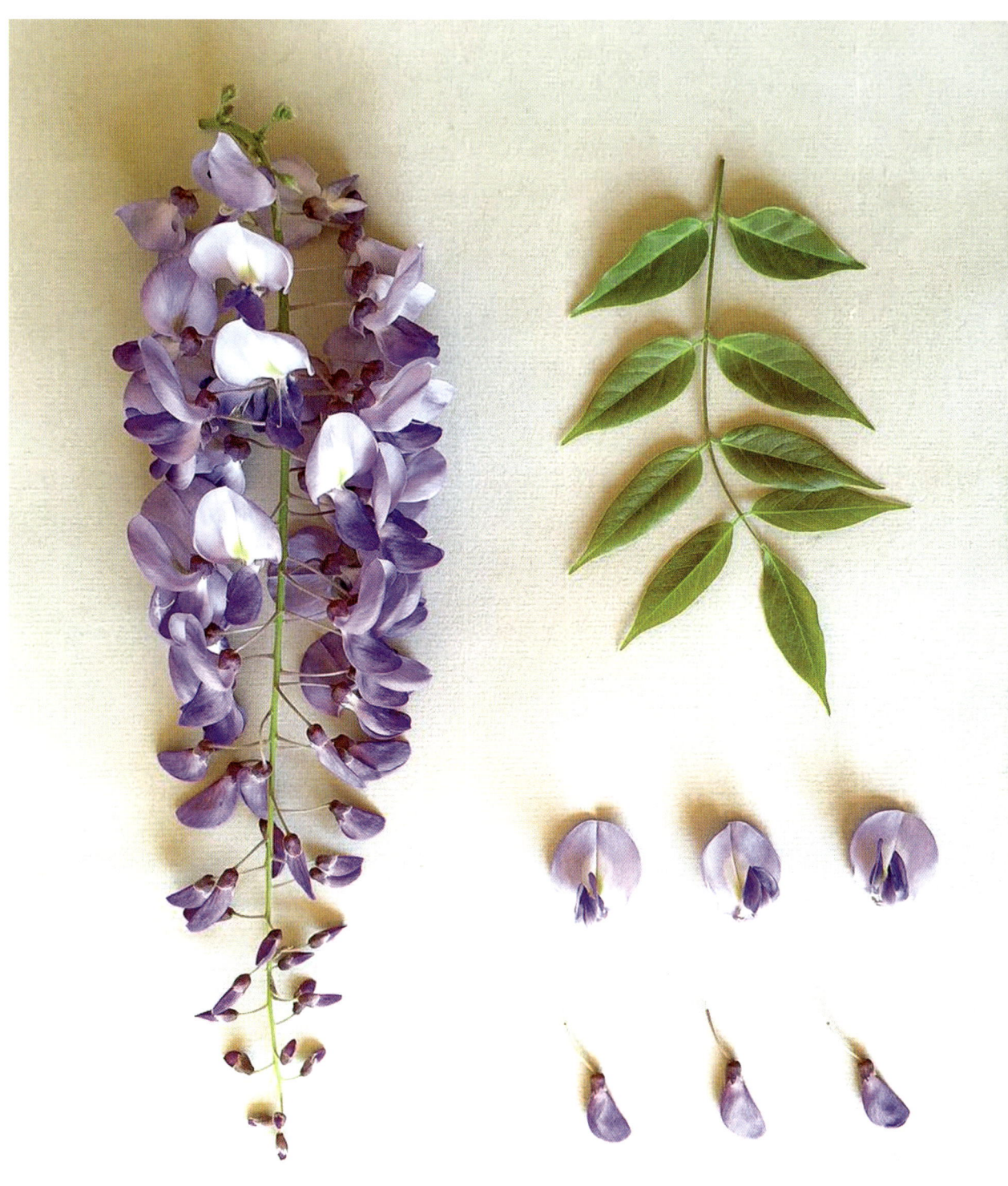

MADRAS LILAC

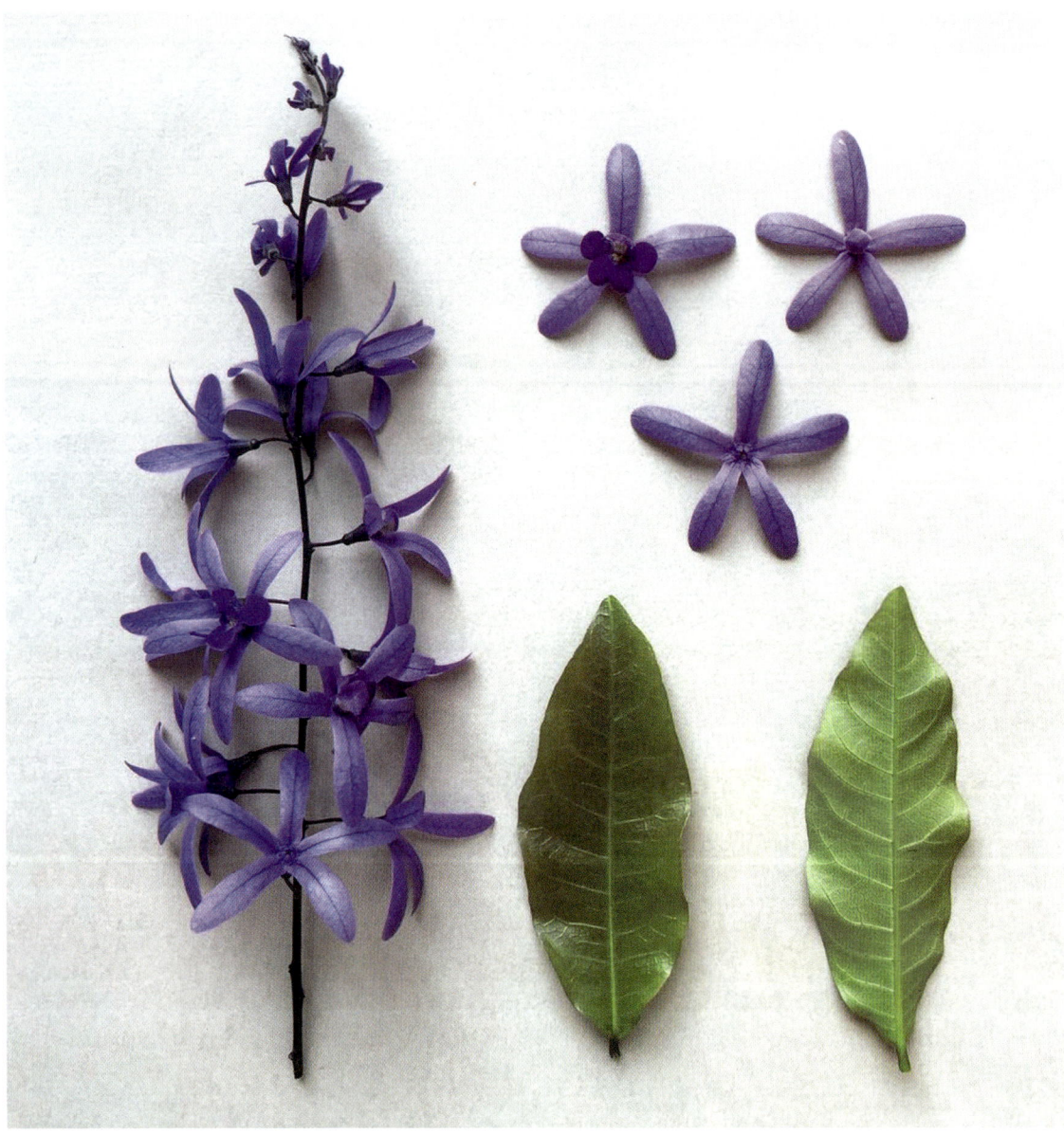

Latin name *Petrea volubilis*
Photograph shows Raceme, florets at different stages of opening, leaves.
Collected Harare, Zimbabwe, 2015
Other common names Sandpaper vine, Queen's wreath, Purple wreath
Family *Verbenaceae*
Origin Woodland from Mexico to tropical Central and South America.
Description A woody, evergreen climber with a twining growth form that can grow up to 6m in height with the aid of a support.
Leaves The ovate-elliptic to elliptical leaves are simple, and arranged in whorls. Their upper surface has a rough texture, hence one of its common names – Sandpaper vine.
Flowers Born in racemes up to 35cm long featuring 15–30 flowers, each made up of five purple sepals surrounding a corolla that is initially purple, but fades to grey.
Fruit Forms winged seeds mainly transported by wind.
Nomenclature Renowned botanist Carl Linnaeus (1707–78) named the genus *Petrea* in honour of a patron of botany, Robert James Petre (1713–42), 8th Baron Petre of Ingatestone Hall, Essex, UK. *Volubilis* is Latin, meaning 'that turns itself around' referring to the twining nature of the plant.
Uses Decorative ornamental.

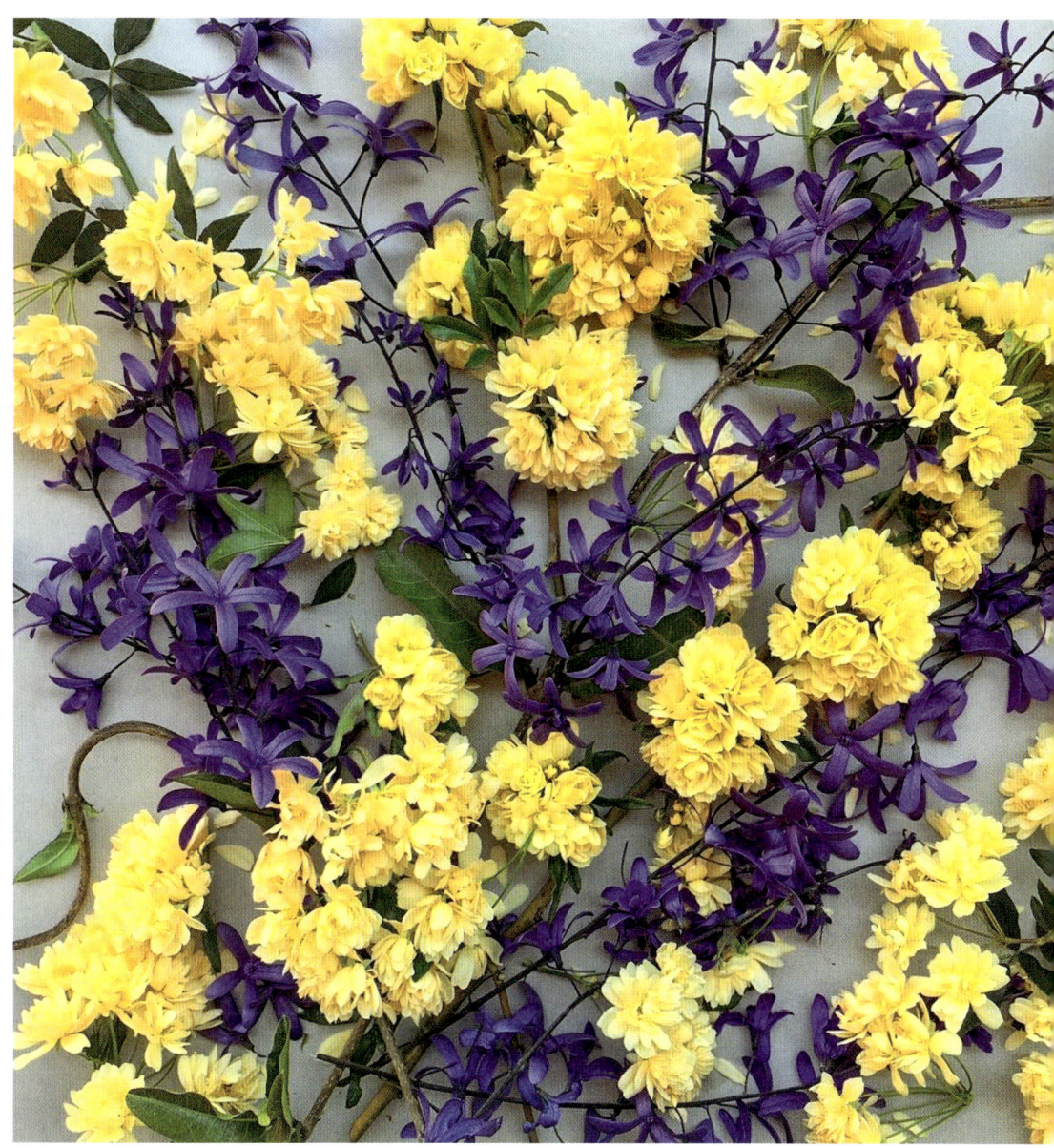

In 1960's Rhodesia the 'de rigeur' garden pairing
was the purple Madras lilac coupled with yellow
Banksia roses, as they bloom at the same time.
Out of fashion and now rarely seen, I love this
nostalgic combination. As children, we called the
vine's twirling, winged seeds, helicopters.

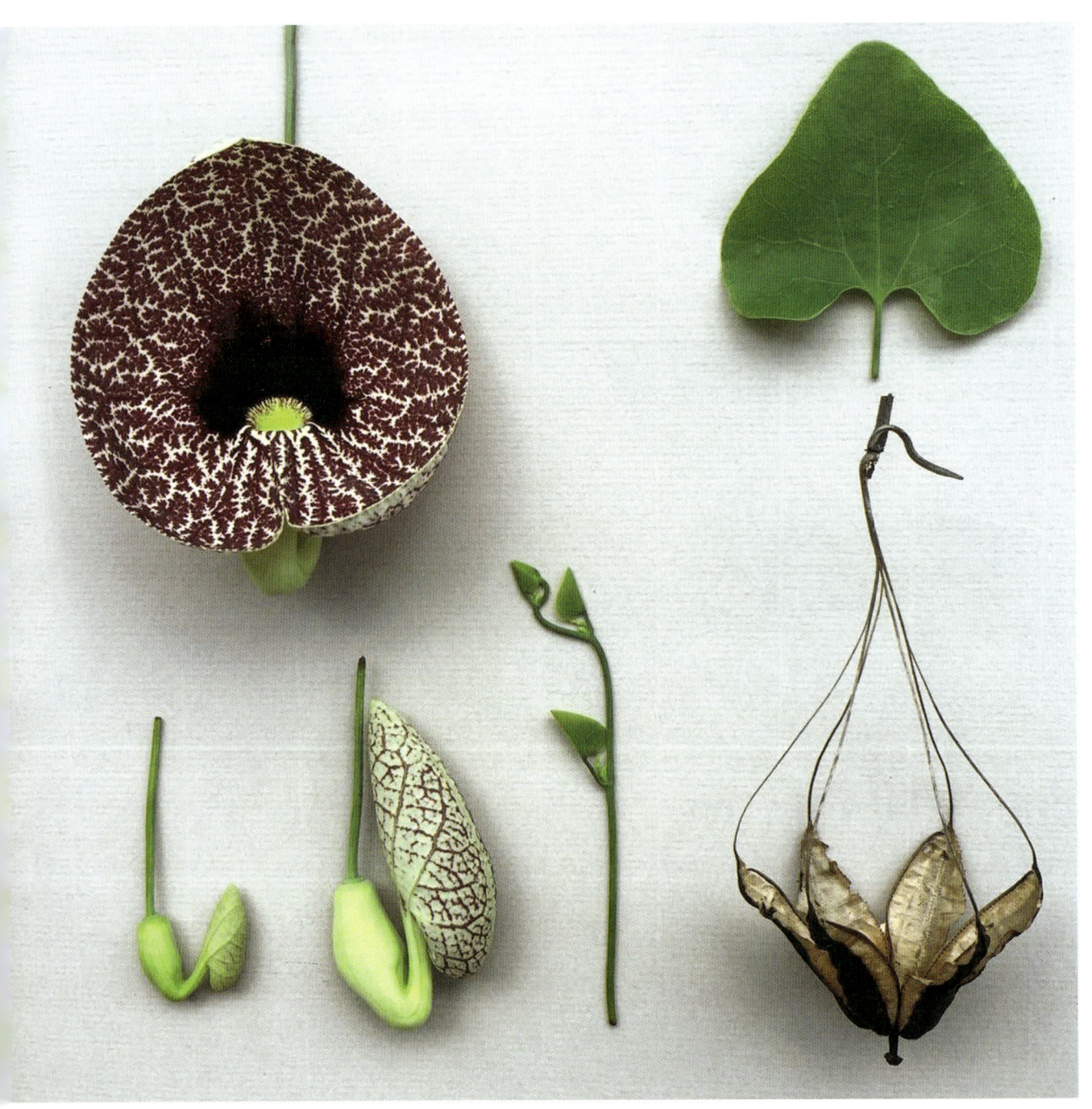

I collected this specimen from La Rochelle, home of Sir Stephen and Lady Virginia Courtauld, who left Europe in the 1950s to build their 'Shangri La' in Africa. The gardens were designed by John Mitchell of London's Kew Gardens, and trees and plants were imported from all over the world.

SCRAMBLING DUTCHMAN'S PIPE

Latin name *Aristolochia littoralis* (syn. *A. elegans*)

Photographs show Opposite: flower, leaf, dried fruit capsule, leaf shoot, mature bud, immature bud.
Right: Fruit, bud, dried fruit capsule, leaf, flowers of *Aristolochia albida*.

Collected Opposite: La Rochelle, Penhalonga, Zimbabwe, 2015. Right: Muchene camp, Zimbabwe, 2016.

Other common name Calico flower, Elegant Dutchman's pipe

Family *Aristolochiaceae*

Origin Temperate and tropical regions of southern and western South America.

Description An evergreen perennial vine that can grow to 5–8m.

Leaves The heart-shaped, grey-green leaves, are 5–10cm across.

Flowers Born in summer, the wonderfully bizarre maroon-and-white, hooded, funnel-shaped flowers are about 10cm across.

Fruit Produces six-ribbed, oblong, cylindrical capsules about 5cm long that remain on the plant. When matured and dry, they open into six segments held together by fibres and resemble a basket or inverted, open parachute.

Nomenclature Genus name *Aristolochia* comes from the Greek words *aristos* meaning 'best' and *lochia*, or 'delivery' (owing to the fact that some species were used as a remedy in childbirth). *Littoralis* is Latin for 'coastline' and *elegans* from its synonym is from Latin, and means 'elegant'.

Uses The common name of a European species *A. clematitis*, which has foetus-shaped buds is 'Birthwort'. As function was believed to follow form, an infusion of this plant was given during childbirth as it was thought to aid the expelling of the placenta. 'Wort' is an old English word meaning a plant. Another African species, *A. Albida*, right, has a bitter root that is made into a tonic to relieve stomach aches. Mixed with lime juice it is a remedy for snake bites and scorpion stings, and an infusion of the dried leaves is used to expel parasitic worms. The leaf, crushed and mixed with castor oil, can be applied topically to pimples.

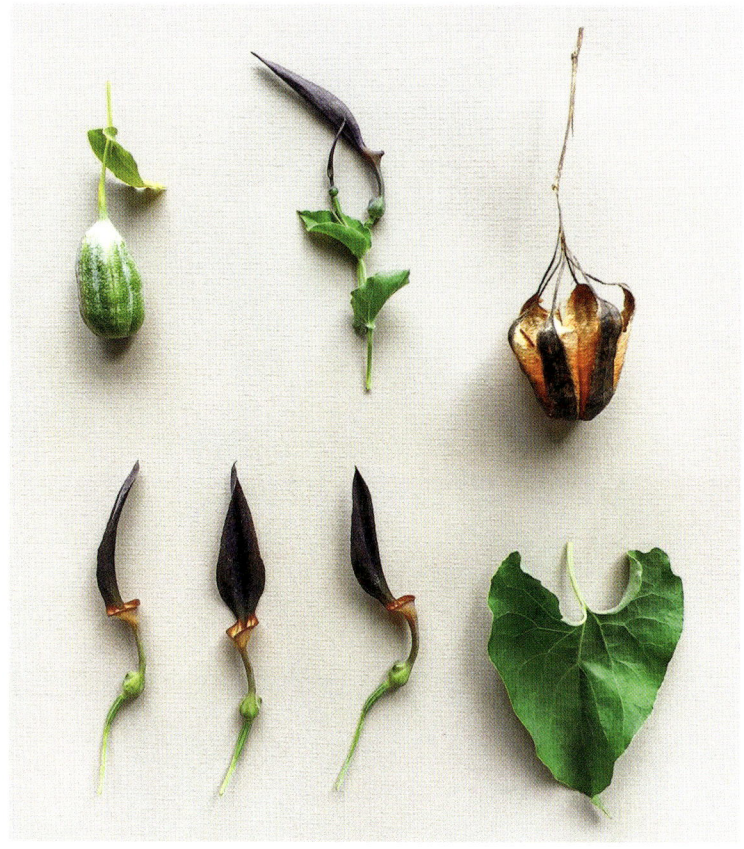

Aristolochia albida, above

This species is native to tropical areas of Africa from Senegal to Ethiopia, south to Angola and Mozambique. I collected it in 2016 on the banks of the Zambezi River near Muchene camp, a place where from the age of ten, I was sent on a series of 'survival courses' that involved canoeing in the river with a maverick guide by the name of Garth Thompson. Ovate to triangular, sometimes three-lobed, leaves are 2–20cm long, hairless to slightly velvety, with three to seven veins from the base. It bears conspicuous leaf-like ovate bracts up to 4cm long on two to 12 flowered axillary, elongated racemes. A grey to purplish-green perianth about 20mm long, extends into a single darker purple lip. It forms oblong to cylindrical, six-ribbed capsules up to 5cm long, that remain on the plant until they split.

PERSIAN LILAC

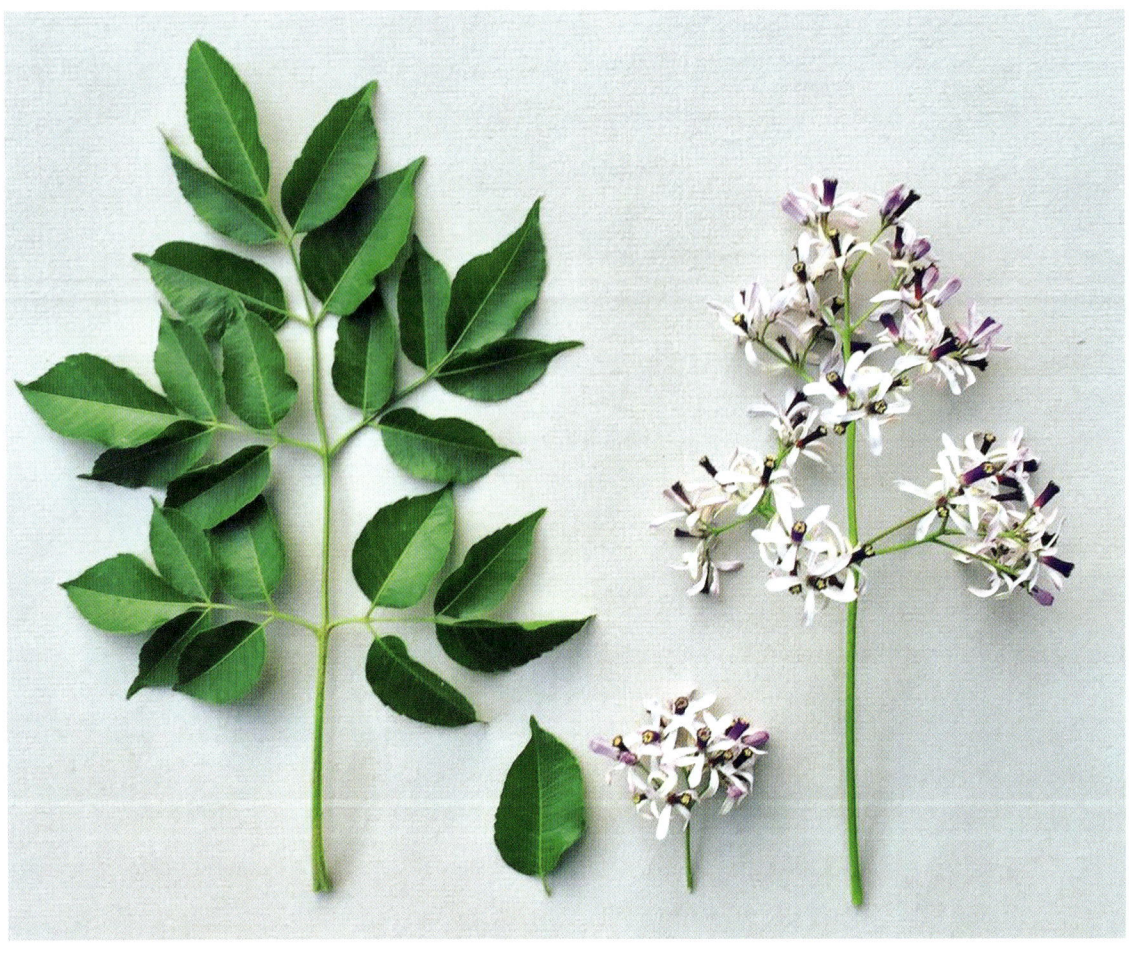

Latin name *Melia azedarach*
Photograph shows Leaves, spray of flowers, cluster of florets, leaflet.
Collected Teviotdale, Harare, Zimbabwe, 2015
Other common names White cedar, Persian lilac, Chinaberry
Family *Meliaceae*
Origin China, Japan, India, South East Asia, northern and eastern Australia.
Description A deciduous shade tree with a rounded crown. This fast-growing tree can be 12m tall with a width of 6-8m at maturity.
Leaves The dark-green, bipinnate leaves have oval to elliptical leaflets, each 2–7cm long.

Flowers Bears clusters of small (2cm across), scented, pale-purple and white flowers, with five petals and stamens that form dark-purple, 6–8mm long, cylindrical 'tubes'.
Fruit Bears small, round (1.5cm in diameter), fleshy berries that are golden yellow when mature and stay on the tree over the winter. Poisonous to humans and some mammals, seeds are eaten by birds who then disperse the seed through their droppings.
Nomenclature In Ancient Greek, *melia* means 'ash tree', *azedarach* is from the Persian for 'noble tree'.
Uses Decorative ornamental. Extracts of the leaves and seeds can be used

to control insects and mites and the termite-resistant timber is made into agricultural implements. Although highly toxic, the tree is used in Chinese medicine for a variety of conditions.

My grandfather's gardener, Witness, affectionately known as Witty, carved catapults from Mulberry wood for me and my brother. The dried, hardened berries of *Melia azedarach* made excellent ammunition.

EMPRESS TREE

Latin name *Paulownia tomentosa*
Photograph shows Stamen, buds, section of panicle with buds, flowers.
Collected St John's School, Harare, Zimbabwe, 2015
Other common names Foxglove tree, Princess tree
Family *Paulowniaceae*
Origin Central and western China.
Description A deciduous tree that grows up to 25m tall, famous for its profusion of flowers in spring.
Leaves Large, opposite, heart-shaped, light-green, velvety leaves can be 15–40cm across.
Flowers Hairy buds develop into fragrant, funnel-shaped, pale-violet flowers, with a cream interior speckled with dark purple. Each flower is 4–6cm long and they are borne on 10–30cm panicles.
Fruit Forms oval capsules 25–38mm long, filled with numerous small seeds. The capsules are initially sticky and green, and later turn brown and dry as they mature.
Nomenclature Named in honour of Anna Pavlovna (1795–1865), queen consort of The Netherlands, and daughter of Tsar Paul I of Russia. *Tomentosa* means 'covered in hairs', referring to the tree's glabrous buds and leaves.
Uses Decorative ornamental. The flowers are edible. The wood is both light and exceptionally strong. Traditionally when Japanese couples had a baby daughter, they planted a *Paulownia* tree. Later when she was ready to marry, the tree was felled to make a dowry chest for her. The flower is also the symbol of the Prime Minister's office in Japan.

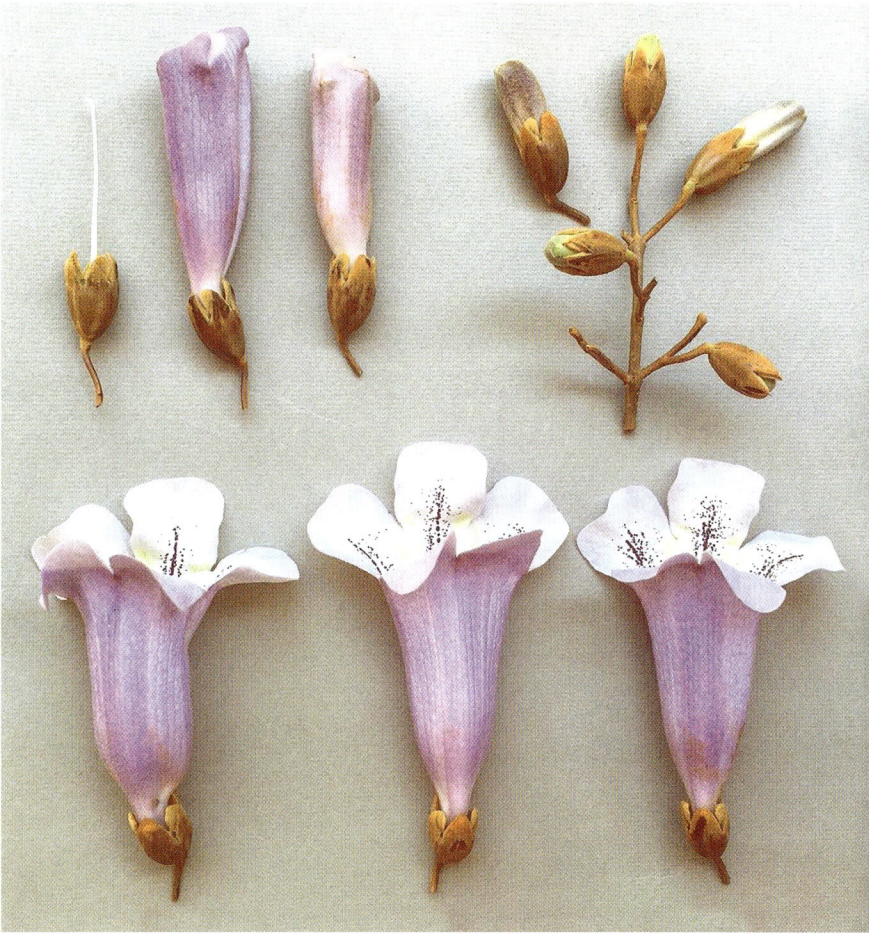

I first encountered a *Paulownia* many years ago while driving in London, UK, with a fellow Zimbabwean, Kimball Groves. Spotting an unknown tree wreathed in purple flowers growing in a garden (which I now know is the Chelsea Physic garden), we pulled over and picked up some of the fallen ones, which we discovered smelled nostalgically like the cough sweets sold in school tuck shops in Zimbabwe. An immediate on-line identification was not possible in those days, and we nicknamed the tree the 'English Jacaranda' as the flowers resembled the jacaranda that line the streets of Harare. Nearly 20 years later, when either of us spies a *Paulownia* in bloom we message the other to say 'The English Jacarandas are flowering'.

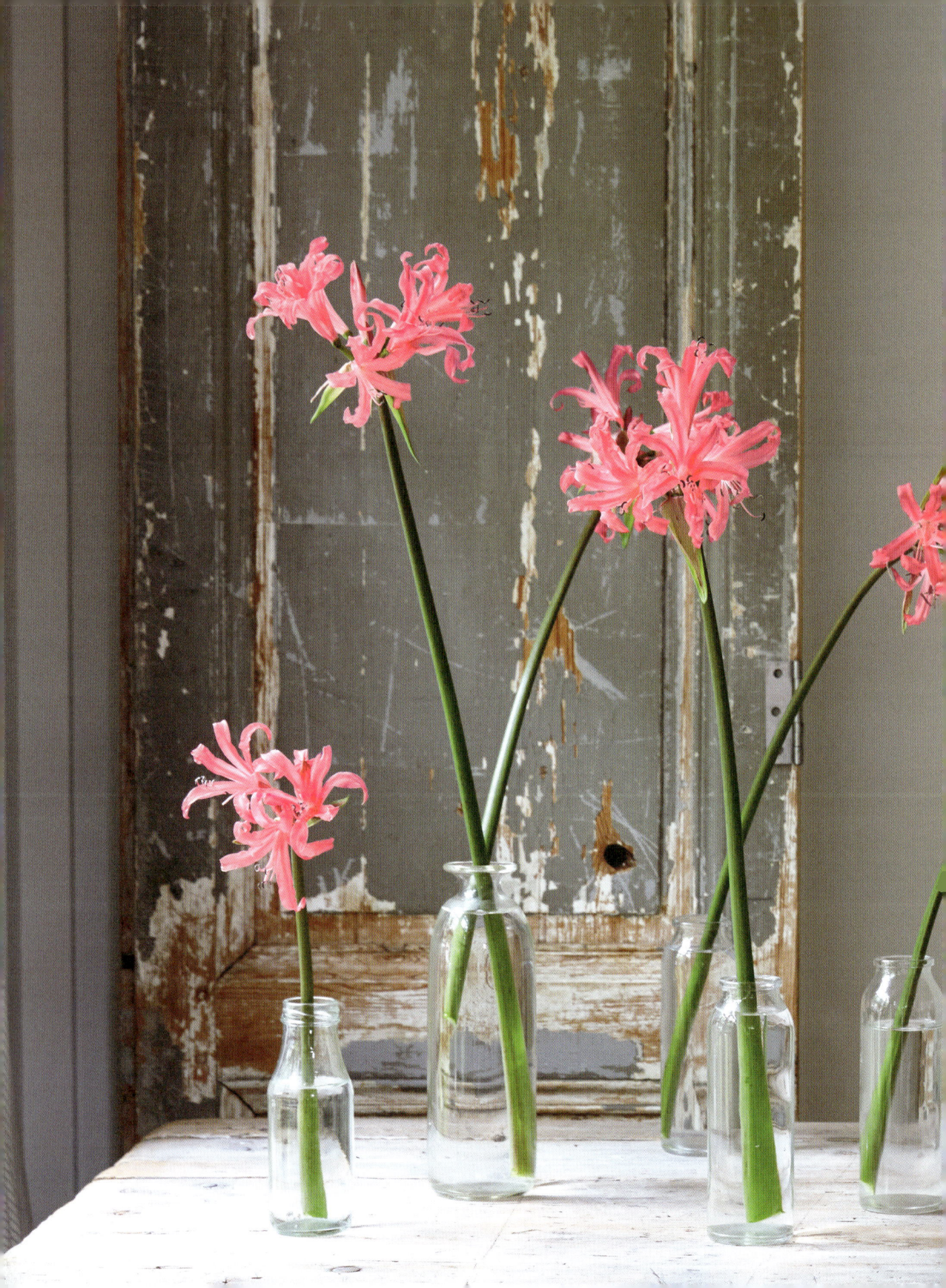

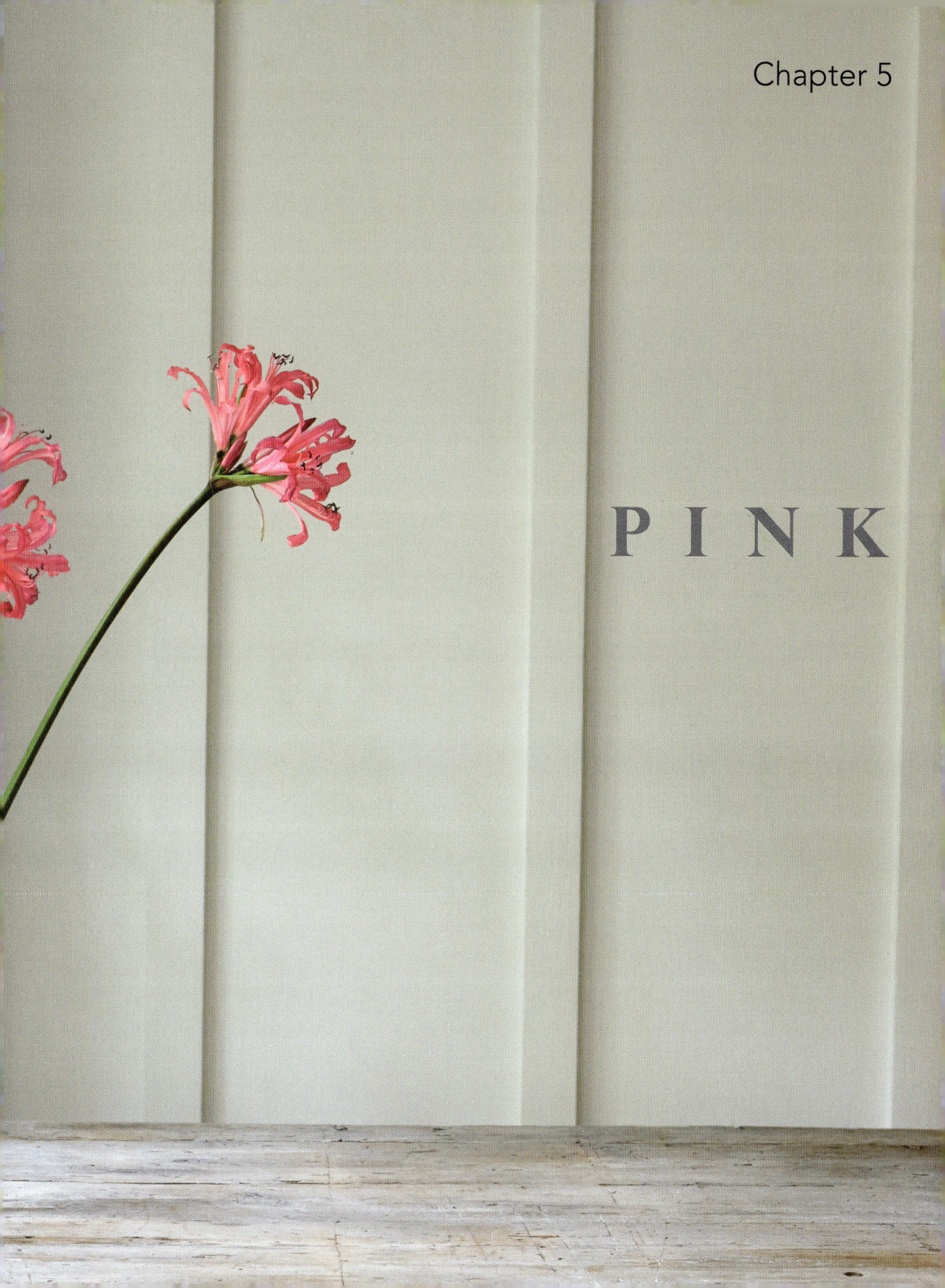

Chapter 5

PINK

SWEET PEA

Latin name *Lathyrus odoratus*
Photograph shows Vine with tendrils, flower buds, tendril, flower, flowers on a vine, keel petal, banner petal, wing petals, stamens with pistil.
Collected Harare, Zimbabwe, 2016
Family *Fabaceae*
Origin Sicily, southern Italy, Cyprus, Aegean Islands.
Description An annual self-supporting climber with winged stems that can reach 2.4m in height, known for its fragrant flowers and intense colours.
Leaves Mid- to dark-green leaves appear in pairs of ovate to elliptic leaflets 5–6cm long. Tendrils form at the base of the bifoliate leaf.

Flowers Stems bear many racemes of two to four large, extremely fragrant flowers that grow up to 5cm long from summer to early autumn. Many cultivars have been developed in a variety of colours. The only colour that is not yet possible is yellow.
Fruit Forms green, hairy pods that turn brown as they mature.
Nomenclature The genus name *Lathyrus* comes from the Greek word *lathyros* (the prefix *la-* means 'very', and the suffix *-thyros* meaning 'passionate'. *Odoratus* is derived from the Latin, *odor* meaning 'the smell' or 'fragrant', refers to the flowers.
Uses Decorative ornamental.

As I write, I have a vase of sweet peas in all shades of amethyst on my desk that were picked early this morning by my mother. Of all the flowers that provoke nostalgia, sweet peas seem to be the ones that most conjure up childhood memories of times spent in the garden, often with my grandparents. Indeed, their scent reminds me particularly of my great grandmother, Jessamine, a woman who adored these flowers and grew them each year. More recently my mother (now a grandmother herself) has started growing sweet peas and I have enjoyed spending time with her each morning pinching tendrils off the vines and collecting the daily crop. Joyfully I have watched my niece Molly join in the picking with her granny, so laying down another generation of familial olfactory memories.

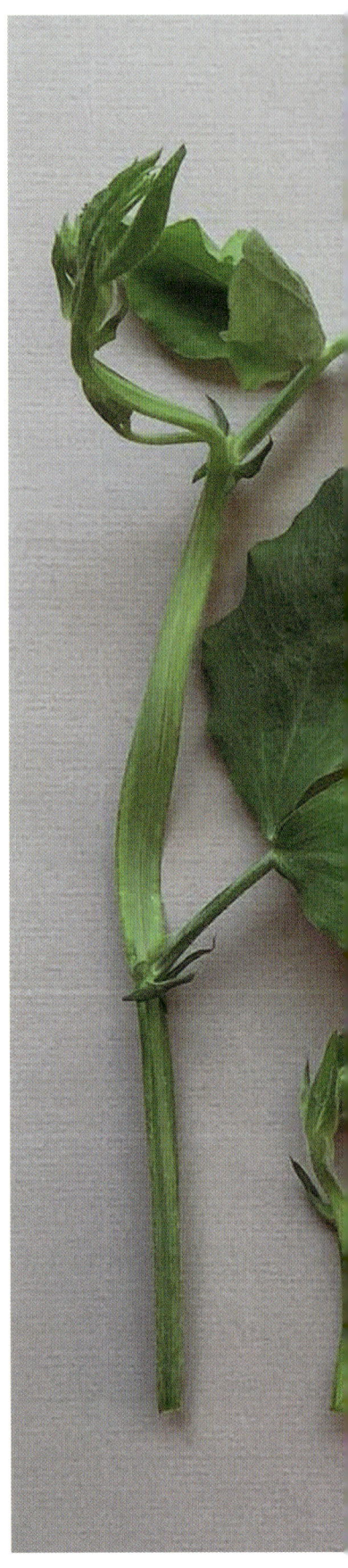

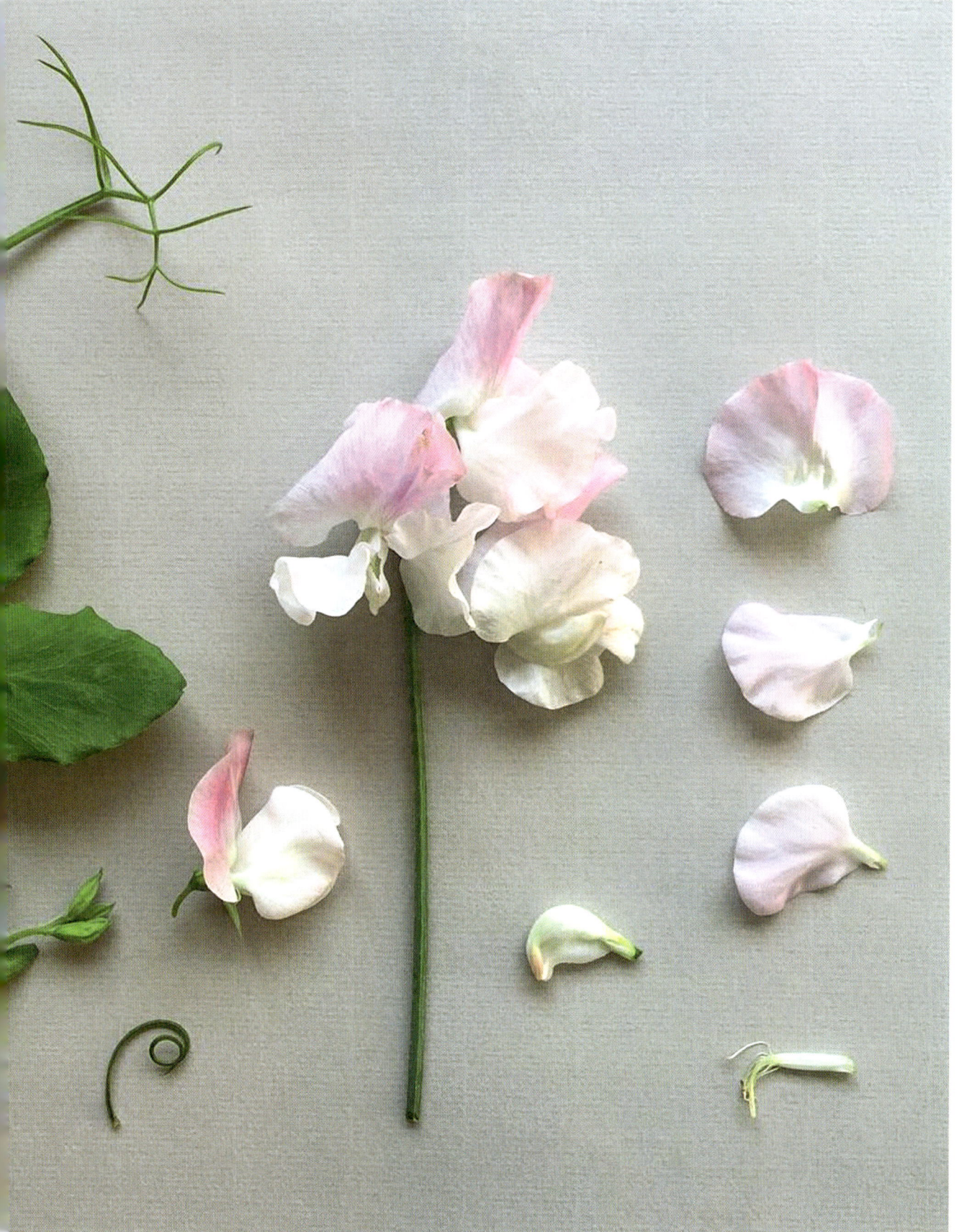

BAMBOO ORCHID

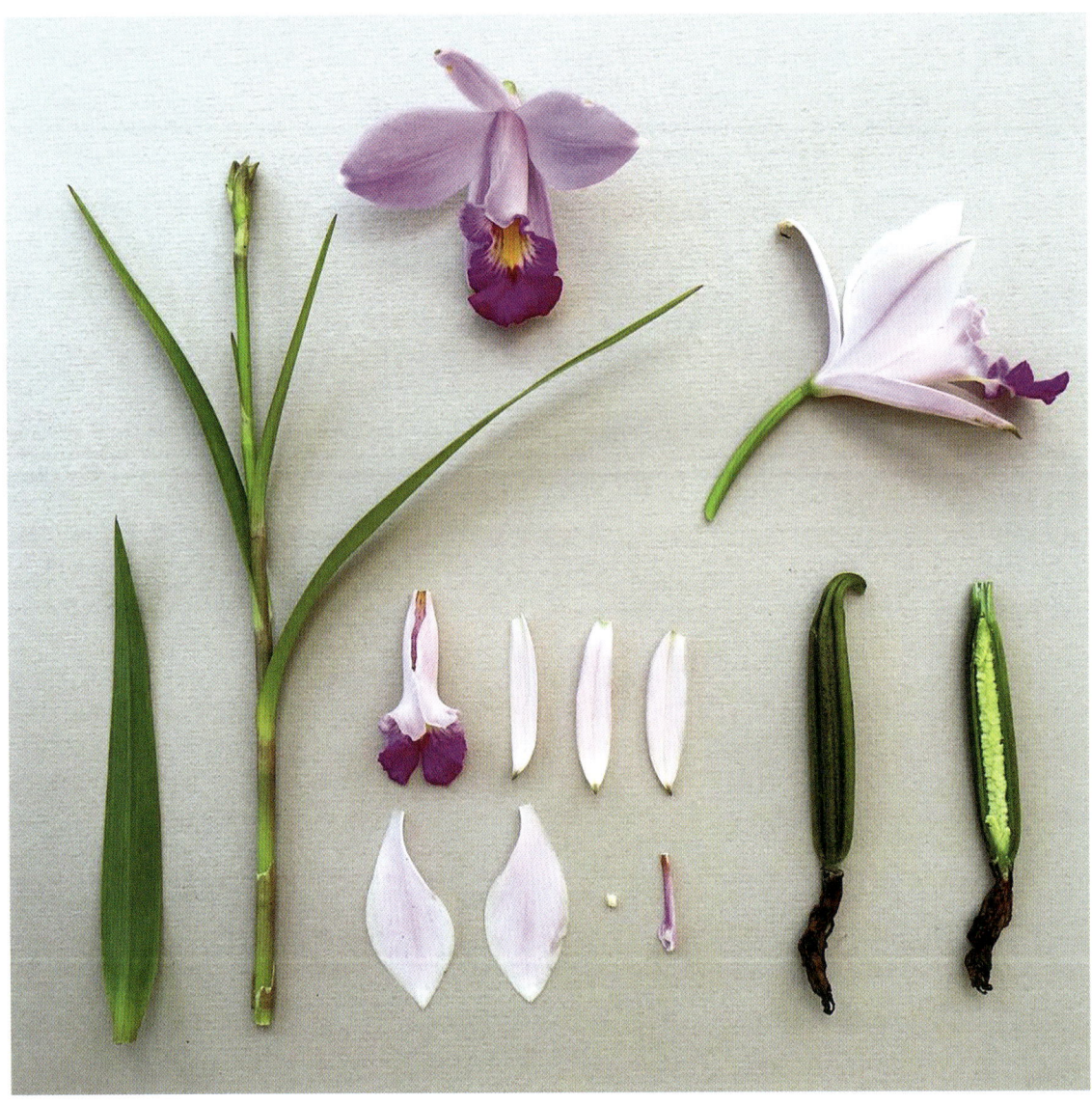

Latin name *Arundina gramanifolia*
Photograph shows Leaf, stem, flowers from above and side, open seed capsule showing seeds, seed capsule, column, anther cap, petals, labellum, sepals.
Collected Trianon, Mauritius, 2017
Family Orchidaceae
Origin Tropical and subtropical Asia from the Himalayas to Tahiti.
Description A terrestrial orchid with foliage that resembles bamboo and forms large clumps. Its reedy stems can be 70–100cm tall.
Leaves Forms alternate, oblong, grass-like, lanceolate leaves that can be 9–19cm long.

Flowers Bears clusters of up to ten flowers (5–8cm in diameter) from summer to autumn. They are white with purple-pink 'lips', similar to those of the Cattleya orchid.
Fruit Produces a cigar-shaped capsule that splits to release the tiny, wind-dispersed seeds when ripe.
Nomenclature The genus name *Arundina* comes from the Latin *arundo* meaning 'reed'/'arrow'. The specific epithet means 'having grass-like leaves'.
Uses Decorative ornamental. In eastern Malaysia, the edible flowers are often stir-fried; they are said to help control high blood pressure.

In Mauritius I spent a day barefoot hiking in the Black River Gorge, with a Namibian friend Severin, and collected this flower, later identified a non-native, invasive species. We followed soft forest paths, then went 'off road' to find a river pool. On our return, we found ourselves on a rocky road, with no footwear or map, but made it back to base with very sore feet.

WATER LILY

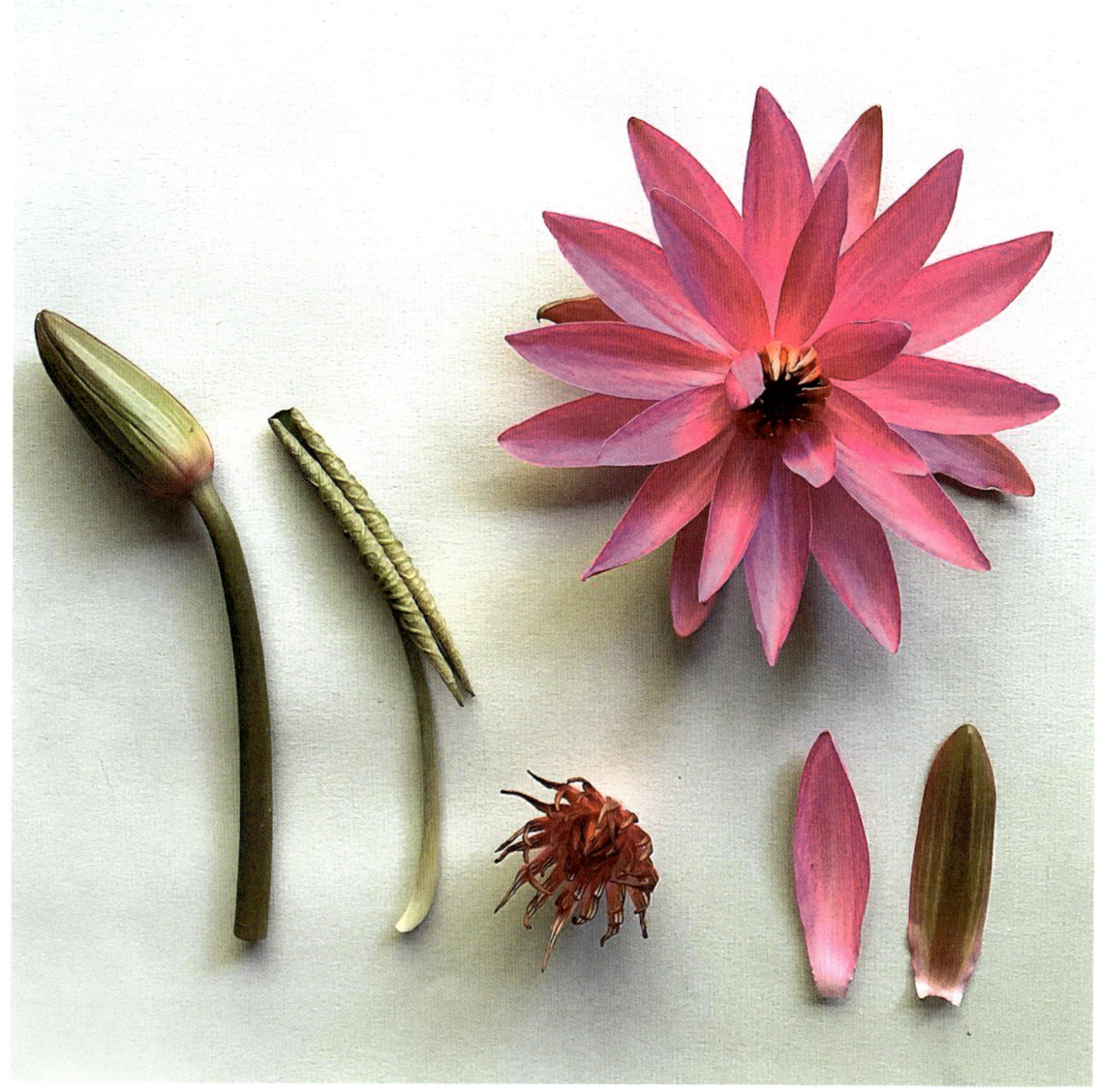

Latin name *Nuphar lutea* 'rubra' (syn. *Nymphaea lutea*)
Photograph shows Bud, rolled leaf, flower, petal, sepal, stamens.
Collected The Datai Hotel, Langkawi, Malaysia, 2016
Other common name Red water lily
Family *Nymphaeaceae*
Origin India
Description Aquatic plant that flowers all year round, widely used for ornamental purposes, as well as a food source to local communities in India.
Leaves Peltate, circular leaves have sharp-toothed margins and can be up to 40cm in diameter. They float on the water surface.

Flowers Bears intense red or rose flowers about 5–6cm across, which are held just above the surface of the water. They have four sepals and many petals and open at dawn and always close at dusk.
Fruit Produces greenish-brown, round berries, that contain numerous small, globular seeds.
Nomenclature Genus name *Nymphaea*, was inspired by the nymphs of Greek and Latin mythology. Cultivar name *Rubra*, meaning red, refers to colour of the flowers.
Uses Decorative ornamental. Young flower buds, stems, and seeds are all edible; in India the seeds are often fried.

On arrival at The Datai Hotel on the Langkawi Peninsular you step into a courtyard where tourmaline-pink water lilies float atop a pond. Sitting on the hotel terrace, sipping iced lemongrass, you can enjoy the magnificent view of the rainforest as it tumbles down to the spectacular translucent, jade-coloured Andaman sea in the distance.

LOTUS

Latin name *Nelumbo nucifera*
Photographs show Flower and leaf.
Collected Beijing, China, 2013
Other common name Sacred Lotus
Family *Nelumbonaceae*
Origin Western Asia from Iran eastwards to China, Japan, and Australia.
Description A rhizatomous, summer-flowering, perennial, marginal aquatic plant that produces individual leaves and flowers directly from its root system.
Leaves Large, peltate leaves with wavy margins, 60cm in diameter, rise above the water surface.
Flowers Peony-like, double, pink flowers – 35cm in diameter – are born on thick stalks that usually rise several centimetres above the leaves.
Fruit Forms conical structures with seeds, each in its own socket.
Nomenclature *Nelumbo* is a Tamil word meaning 'blue', and *nucifera* is derived from the Latin words, *nux*, or 'nut', and *fera*, meaning 'bearing', and refers to the plant's seeds.
Uses All parts of this plant are edible and a tea is made from dried flower stamens. The flowers and dried seed heads can be cut and used decoratively.

I was a confident botanical smuggler from a young age, having learned from my father to hide such contraband in my wash bag. On a trip to Mauritius as a teenager, I collected lotus seeds from the Pamplemousse Botanical Gardens, and brought them back to Zimbabwe with grand ideas for Maduma's lily pond. After various failed attempts at germination, I found that filing the ends of the seeds down and soaking them in hot water did the trick. But, alas, the resulting plantlets never grew beyond seedlings, so I never became a lotus eater. Last year I found myself back in the Pamplemousse Botanical Gardens again, marvelling at the enormous leaves of the Amazonian water lilies. Once again I collected some lotus seeds and brought them back to Zimbabwe. This time around, with the Internet, I found information telling me how to cultivate the seeds successfully. I now have young lotus plants growing in buckets at Chinakwaremba ready to be transferred to the lily pond at Maduma once they are established.

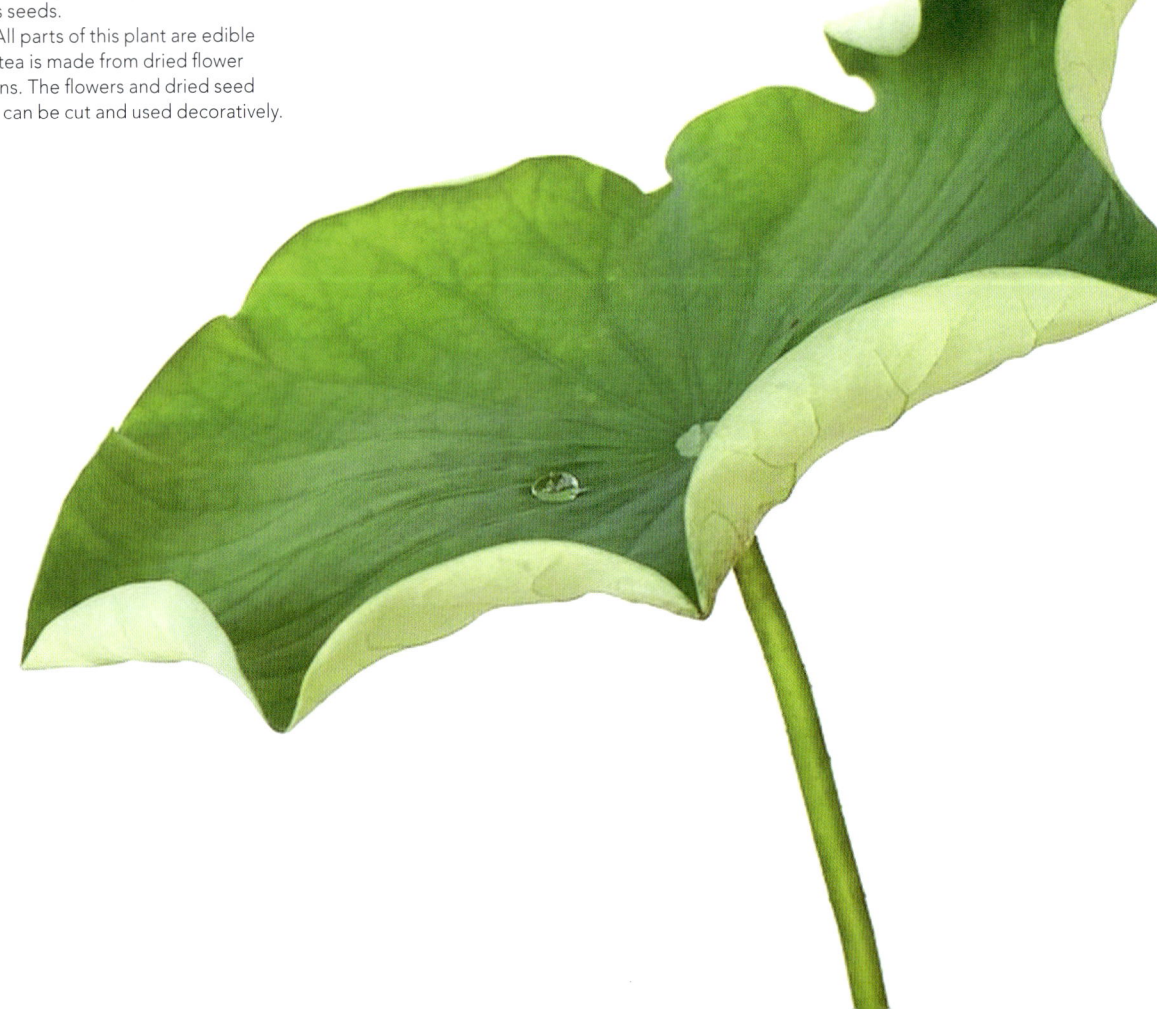

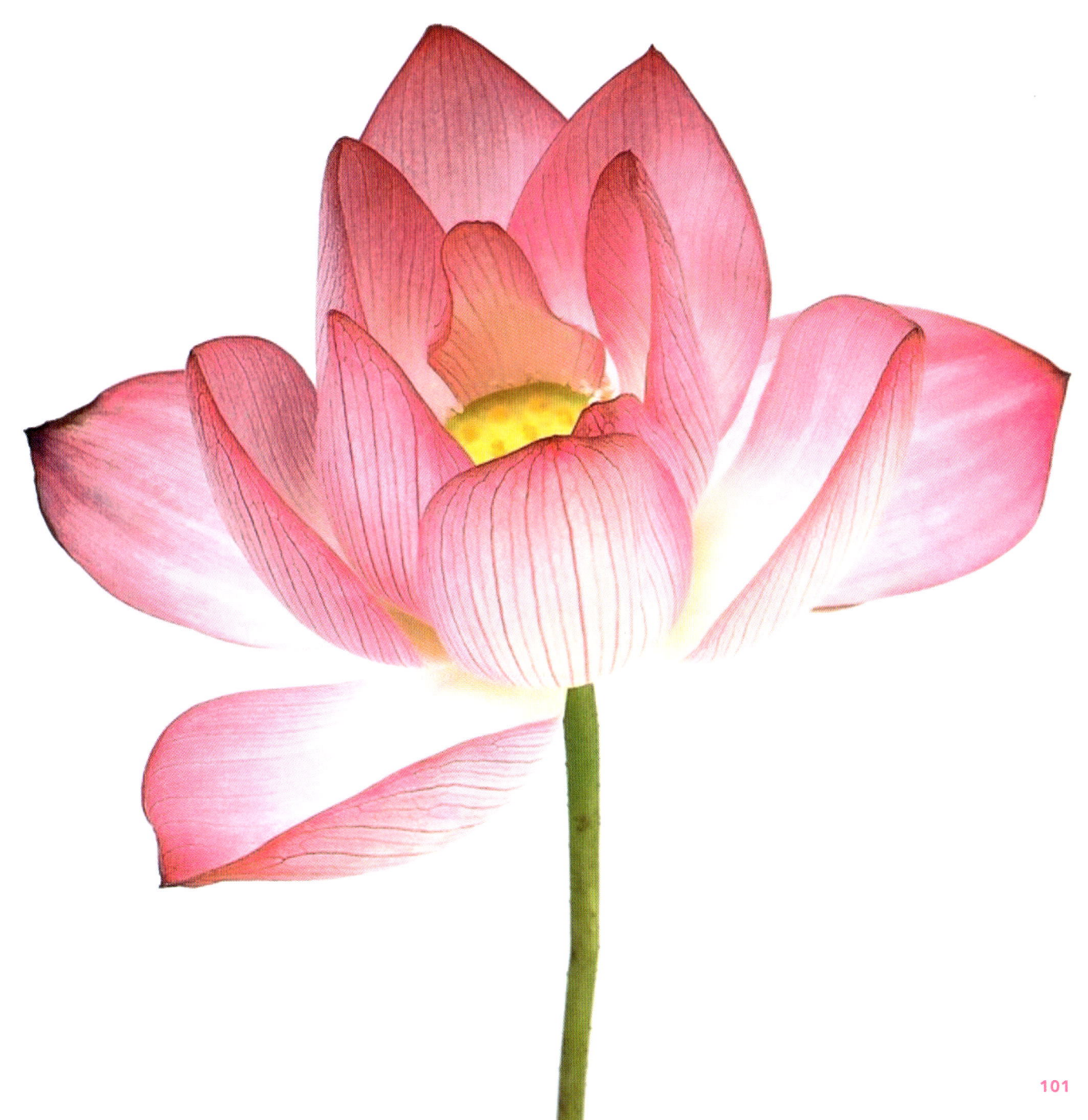

CARRION LILY

Latin name *Stapelia gigantea*
Photograph shows Stem/leaf, bud, open flower with green bottle fly (*Phaecinia sericata*).
Collected Harare, Zimbabwe, 2016
Other common names Giant toad flower, Starfish flower, Zulu giant
Family *Apocynaceae*
Origin Zambia, Malawi, Mozambique, Botswana, Zimbabwe, South Africa.
Description Cactus-like, succulent shrub with shallow root system that grows up to 30cm tall with a spread of 1m.
Leaves Deeply ribbed, green, erect, hairy, stem-like leaves are 2–3cm wide, with obscure stipular glands.
Flowers In summer, the plant bears huge, foul-smelling, star-shaped flowers, 30–45cm in diameter. Pale-yellow with crimson stripes and centre, they have crimson/purple hairs on the undersides and are pollinated by flies.
Fruit Explosive seed pods, produced in pairs and united at base, contain small flat, tufted seeds that are dispersed by wind.
Nomenclature The genus is named in honour of the 17th-century Dutch physician and botanist Johannes van Stapel (1602–36); *gigantea* refers to the size of the flowers.
Uses Botanical interest. This plant is purported to be used as an antidote to hysteria by Zulu people.

A morbid curiosity compels me to get down on my hands and knees and smell the blooms of this plant. The flowers exude an overpowering stench that is reminiscent of rotting flesh.

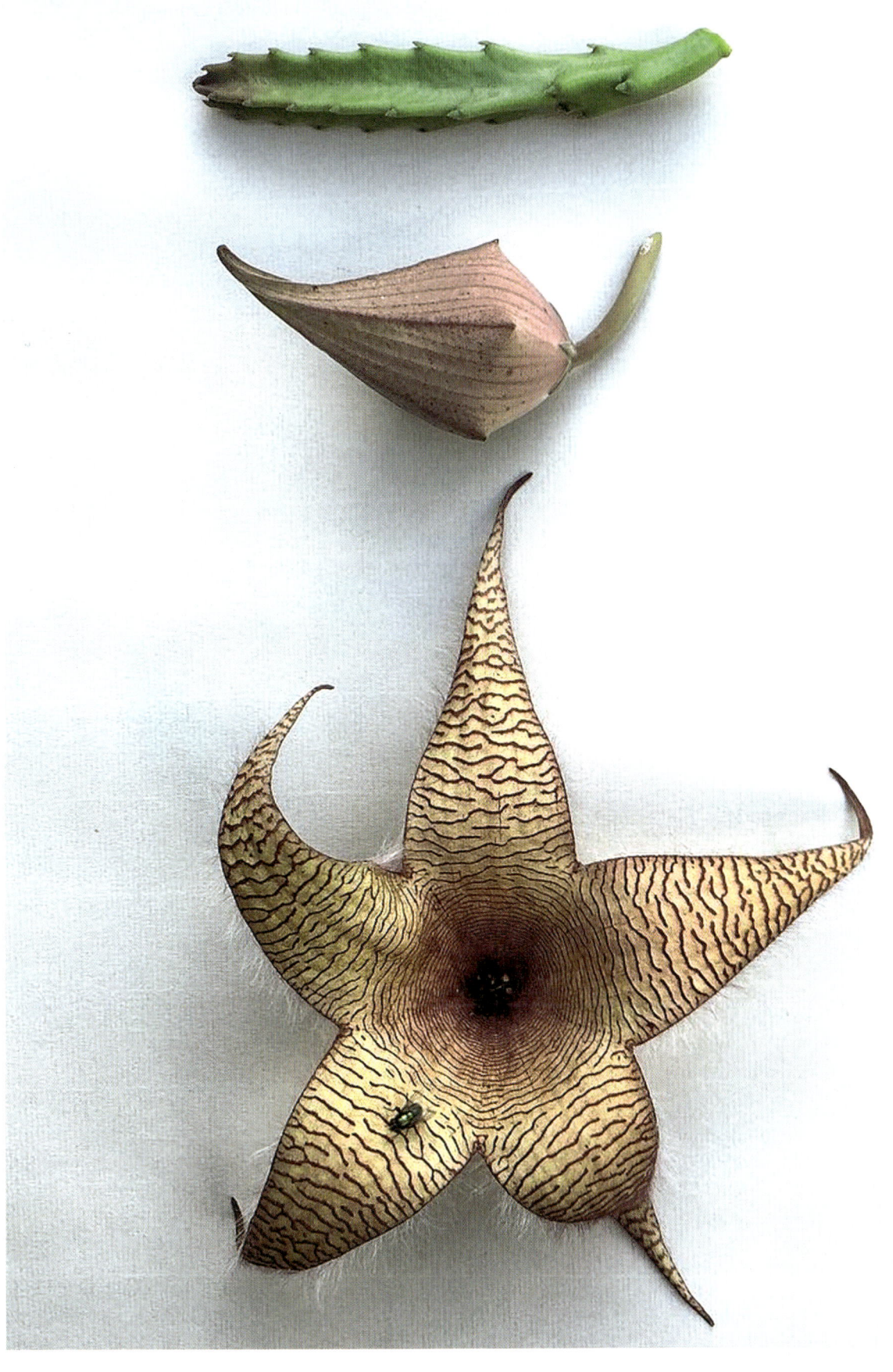

TULIP MAGNOLIA

Latin name *Magnolia* x *soulangeana*
Photograph shows: Buds, carpel, fruit, bud, flower, mature bud, tepals, stamen.
Other common names Chinese magnolia, Saucer magnolia
Family *Magnoliaceae*
Origin Parents of this hybrid – *M. denudata* and *M. liliflora* – are both native to China.
Description A multi-stemmed, deciduous, large shrub or small tree with a height of 8m and a spread of 6–9m that has a bright, attractive, grey bark.
Leaves Obovate and alternate, pubescent leaves grow up to 20cm in length. They are dark green in summer and turn brown in autumn, before shedding in winter.
Flowers In spring, before leaves emerge, the plant bears large, goblet-shaped flowers in various shades of white, pink, and maroon that are commonly 10–20cm in diameter.
Fruit Elongated fruit are about 7cm long, and house small red seeds.
Nomenclature The genus *Magnolia* is named after the French botanist, Pierre Magnol (1638–1715). This hybrid was bred by the French plantsman Étienne Soulange-Bodin (1774–1846) in 1820. Impressed with the resulting progeny, the hybrid quickly entered cultivation.
Use Decorative ornamental.

The Tulip magnolia is one of the most magnificent deciduous flowering trees. Its velvety buds burst forth from silvery naked stems and open into scented, pale pink goblet-shaped flowers long before the flowers. The French horticulturalist who cross bred it, Etienne Soulange-Bodin, was originally a Parisian soldier and *M. soulengeana* and its cultivars were his most successful plants.

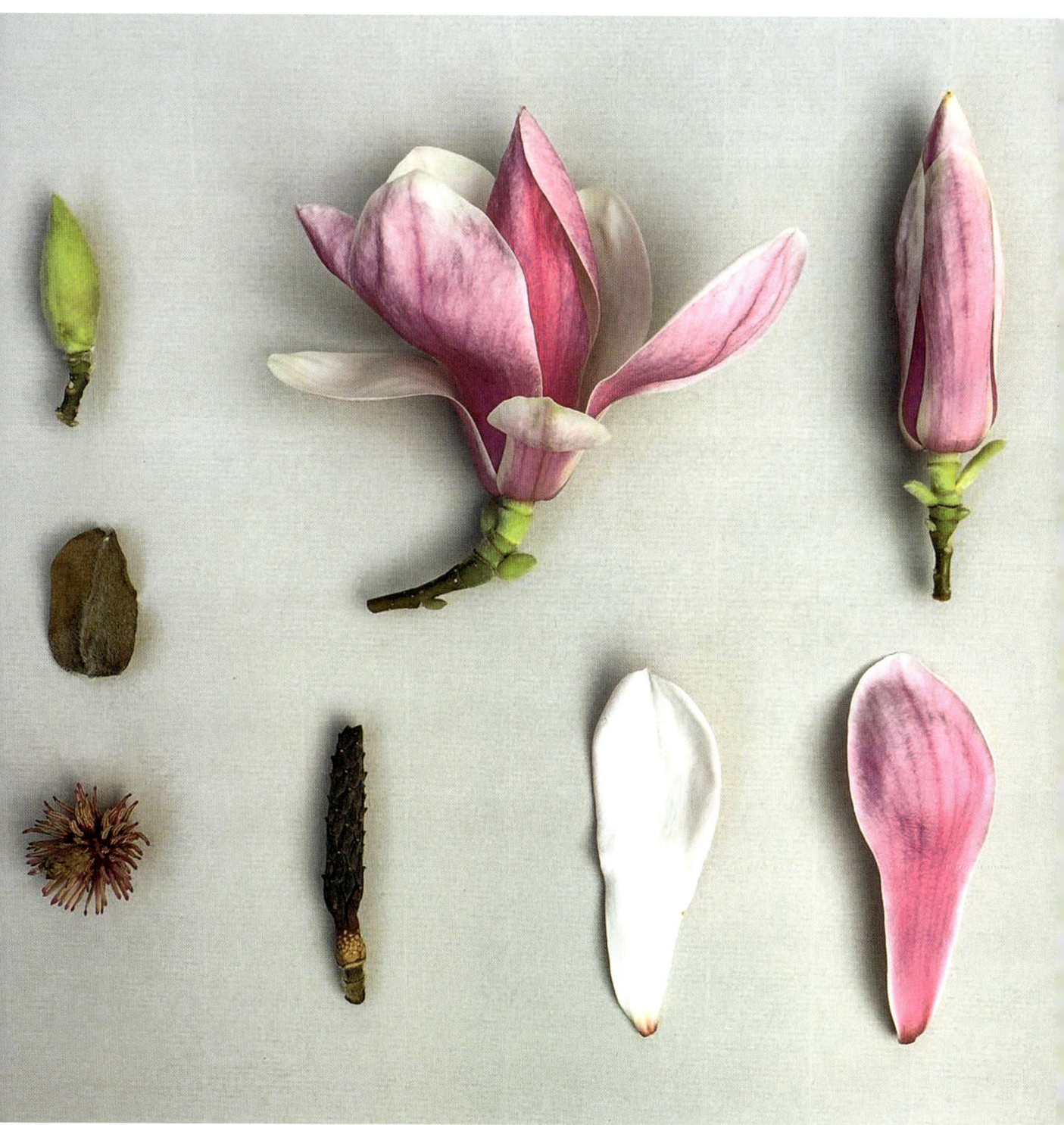

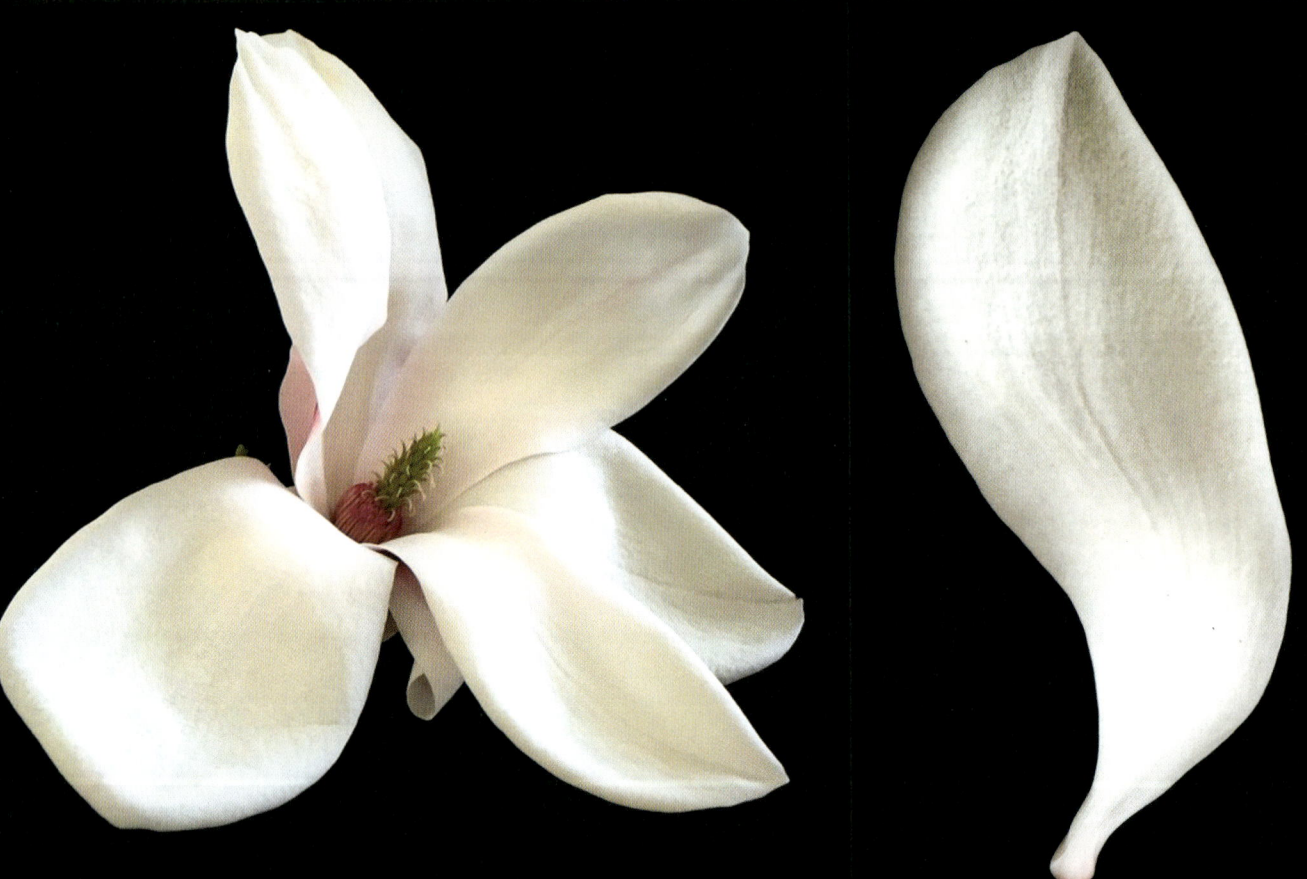

The pale-mauve colour of these
petals reminds me of a pudding
that my grandmother made
which she called 'Stone cream'.
She stirred port into cream
and the resulting mixture was
a wonderful shade of palest
mauve – just like these flowers.

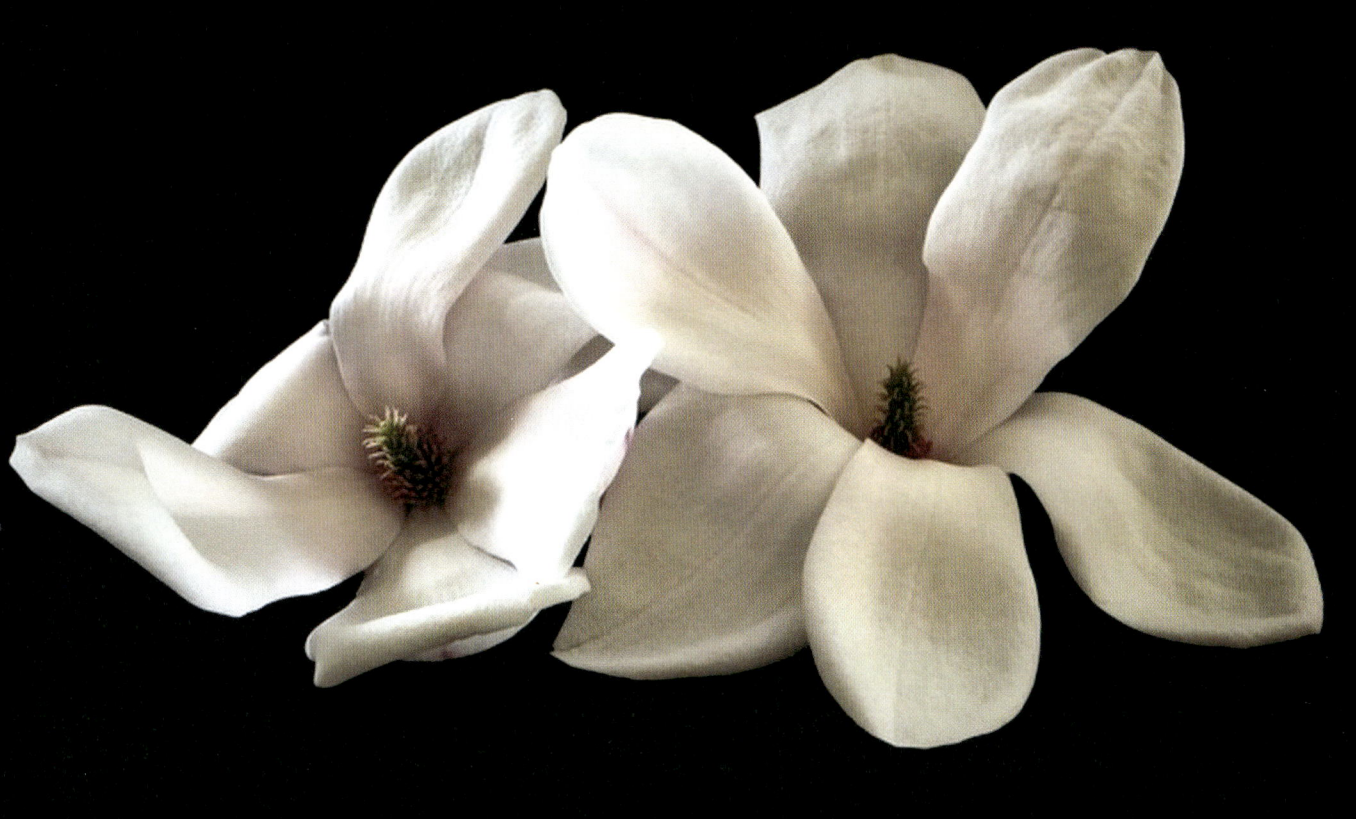

PINK CORAL TREE

Latin name *Erythrina caffra*
Photograph shows Leaf, inflorescence, stamens, flowers.
Collected Harare, Zimbabe, 2016
Other common name Coast coral tree
Family *Fabaceae*
Origin South-eastern Africa
Description A medium to large deciduous tree, with a rounded spreading canopy, of up to 20m, depending on growing conditions.
Leaves The trifoliate leaves are broadly ovate or elliptic; the terminal leaflet grows up to 12cm long, while the lateral ones are smaller.
Flowers Pink, red, and scarlet flowers are borne in dense clusters on the heads of fleshy stalks in the dry season, before leaves develop. When the flowers open, the stamens are exposed, giving the flowers a 'whiskered' appearance. The petals of *E. caffra* are longer, narrower, and closer to the stamen than *E. lysistemon* (page 25).
Fruit Forms dark-brown, narrow, cylindrical seed pods up to 10cm long, which turn red-brown with age, then split when dry to release small coral-red seeds with black spots on one side.
Nomenclature *Erythrina* comes from the Greek word *erythros*, or 'red' describing the flower colour of some of the species. *Caffra* is from Hebrew, and means 'a person living off the land'; the word was often used to describe plants from the eastern parts of South Africa.
Uses Medicinal. The bark is used topically to treat wounds and toothache.

This form of *Erythrina caffra* with shell pink blooms is rarely seen. I have found a few of them in the sea of the red flowering specimens growing in Harare.

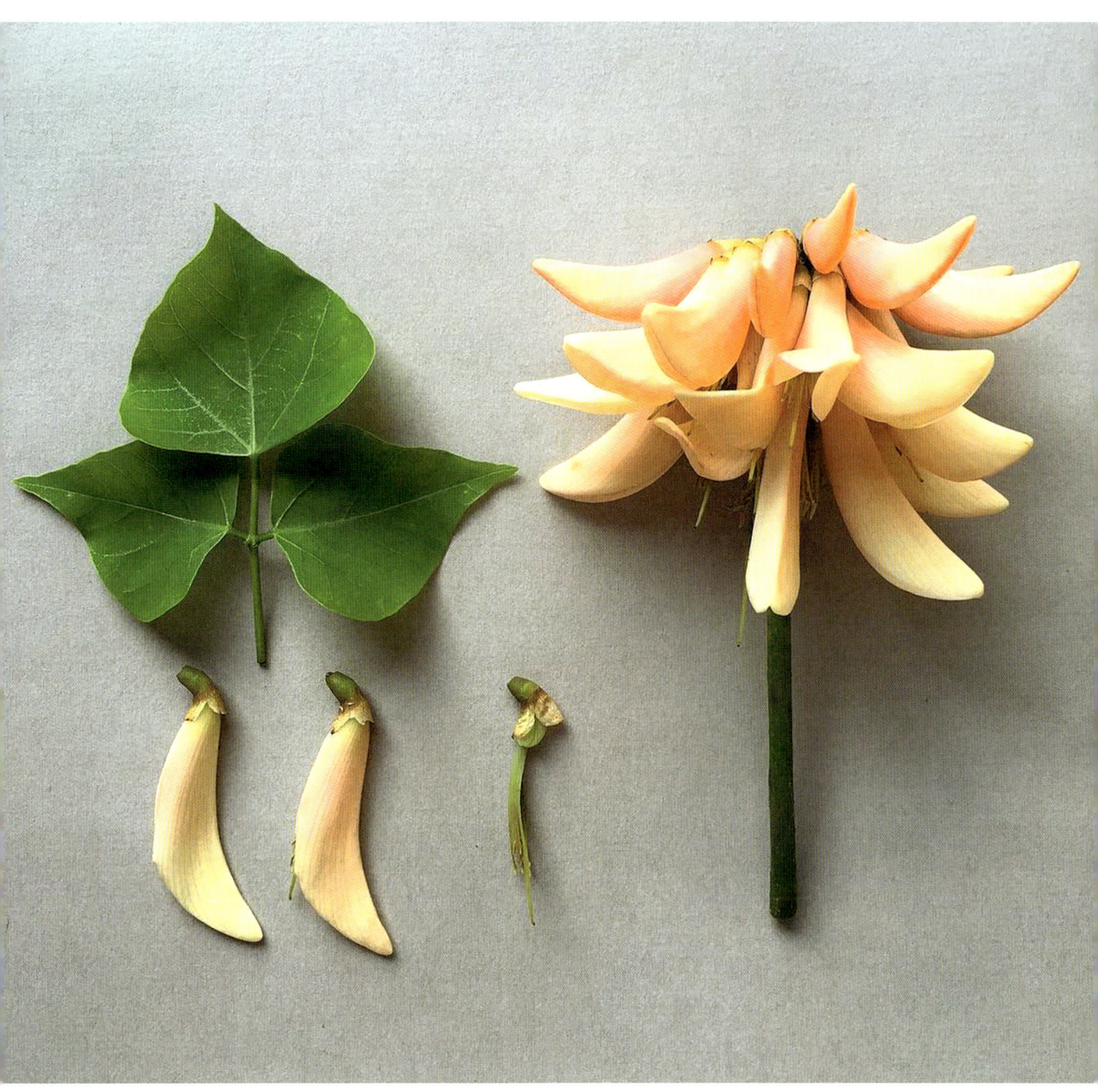

HONG KONG ORCHID TREE

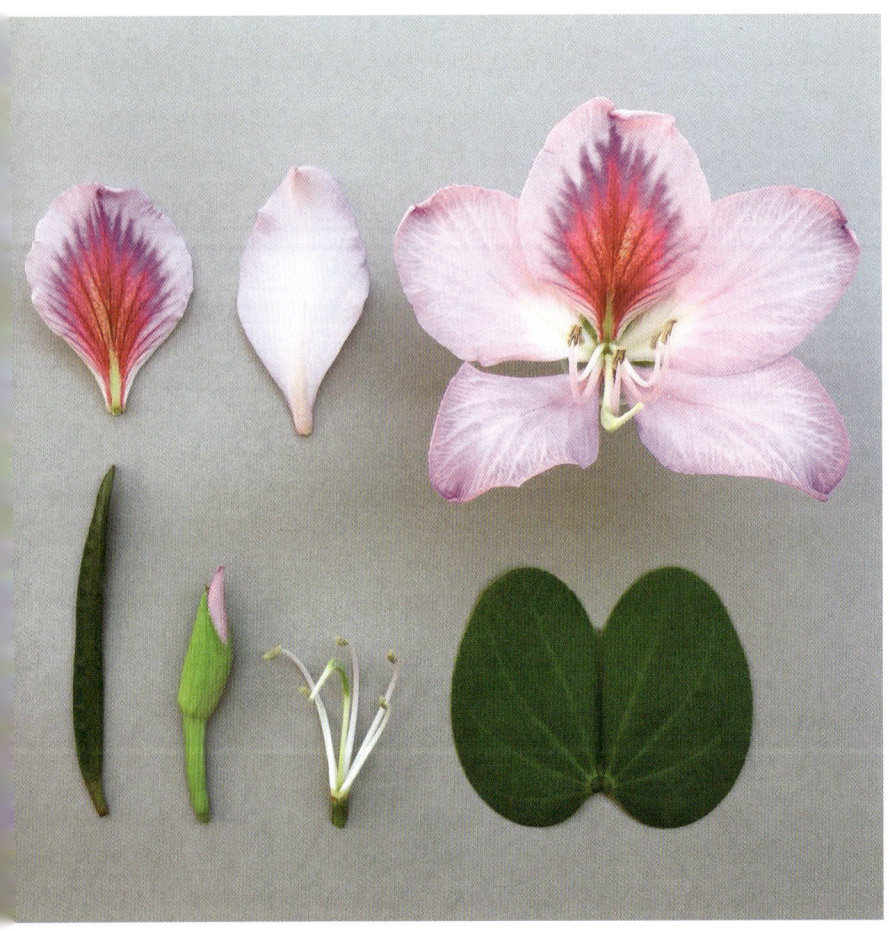

Latin name *Bauhinia* x *blakeana*
Photographs show Left: petals, flower, leaf, stamens, bud, fruit. Opposite: stems with flowers, buds, and petals.
Collected Harare, Zimbabwe, 2015
Other common name Bauhinia tree
Family *Fabaceae*
Origin Hong Kong, southern China.
Description A small, multi-trunked, evergreen to semi-evergreen/semi-deciduous, flowering tree that can be 6–10m in height.
Leaves Bilobed, light-green leaves resemble butterfly wings and are 7–10cm long and 10–13cm wide.
Flowers Bears large (15cm across), fragrant, showy, deep- and light-pink, flowers that have five petals and five or six sterile stamens.
Fruit None
Nomenclature The genus *Bauhinia* is named after Swiss-French botanist brothers Johann (1541–1613) and Gaspard (1560–1624) Bauhin. *Blakeana* is in honour of Sir Henry Blake (1840–1918), British Governor of Hong Kong from 1898 to 1903 and his wife, Lady Blake.
Uses Decorative ornamental tree.

The flower of this tree is depicted on the Hong Kong flag, its coat of arms, and its coins. These trees line the streets of Harare and bloom in a glorious array of pink and white and herald the jacaranda flower season that follows, when the city becomes a sea of purple.

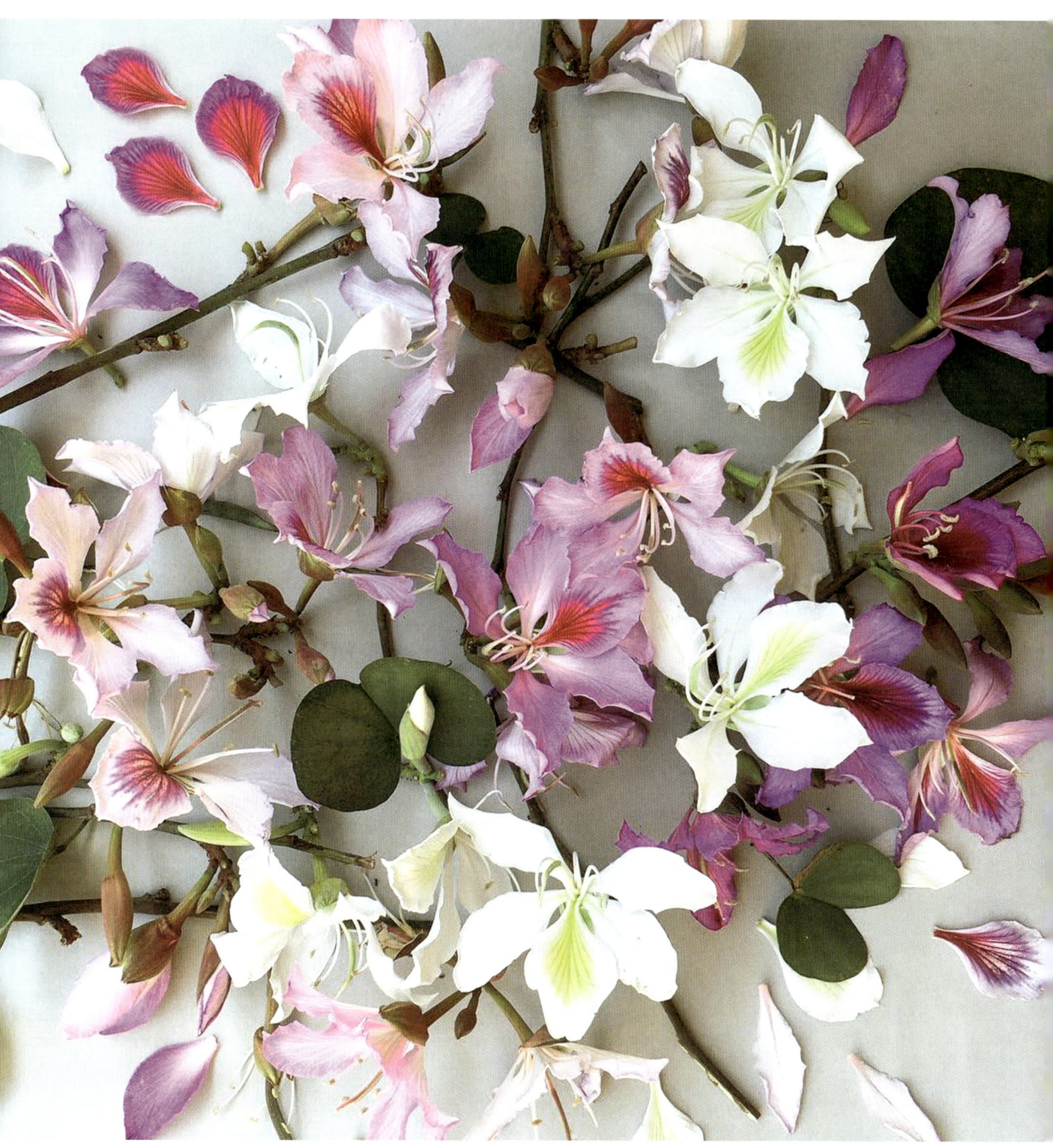

SILK FLOSS TREE

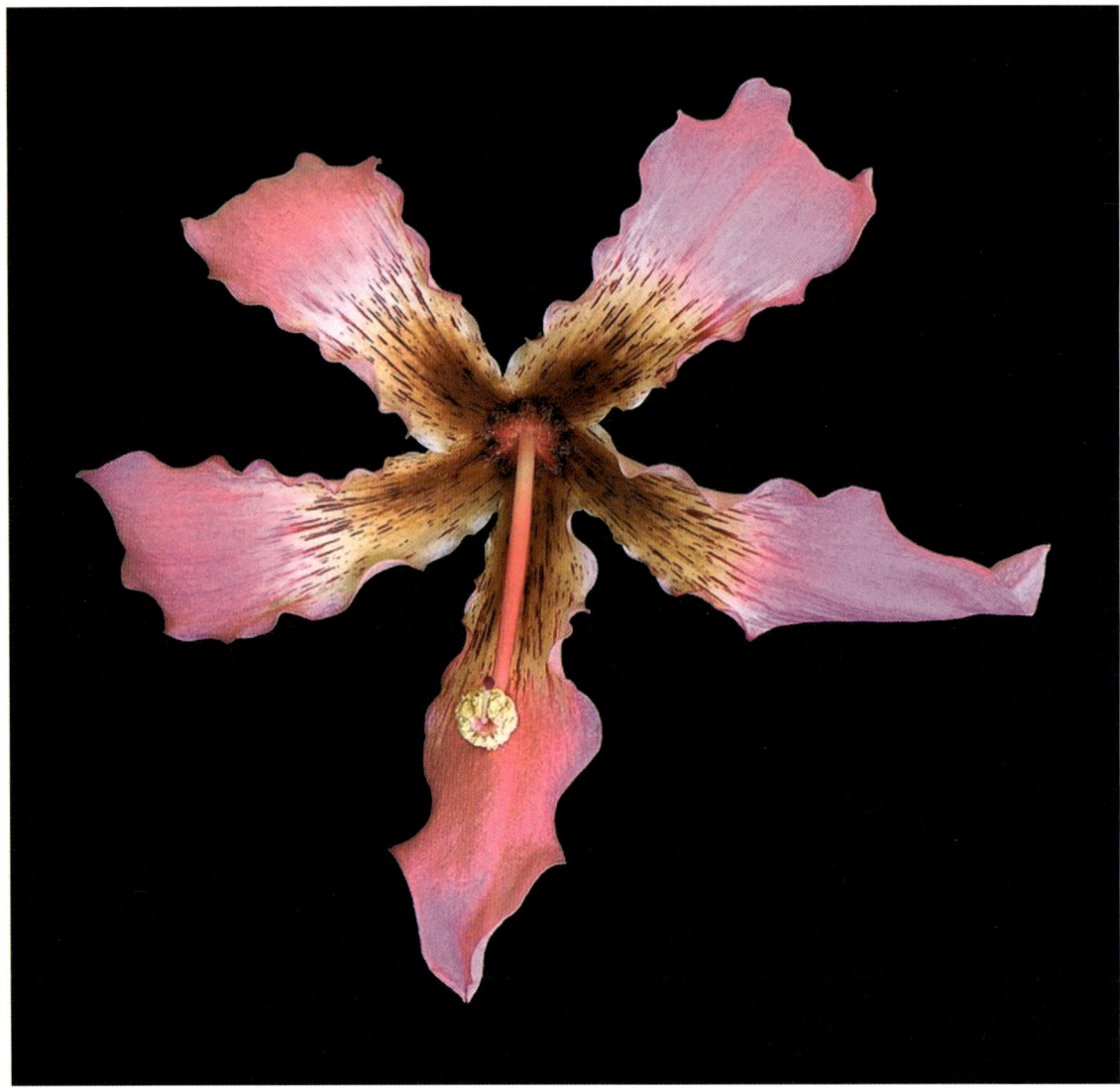

Latin name *Ceiba speciosa* (syn. *Chorisia speciosa*)
Photograph shows Flower
Collected Harare, Zimbabwe, 2015
Other common name Floss silk tree, Kapok tree
Family *Malvaceae* (formerly *Bombaceae*)
Origin Tropical and subtropical forests of South America.
Description A large deciduous, autumn to winter flowering tree with wide, spreading branches that grows up to 30m. When young, the trunk is green and covered with warty triangular spines; trunk usually turns grey and spineless as it ages. The young trees are straight and narrow, then slowly develop broad, spreading, umbrella canopies.

Leaves Palmate-compound leaves consist of five to seven pointed leaflets, each up to 10 cm long and with serrated margins, that drop just before flowering.
Flowers Large, showy pink to purple flowers with white centres and brown bases have five petals, and grow up to 15cm in diameter.
Fruit Forms large, woody, pear-shaped capsules up to 20cm long, containing many black, pea-like seeds, enclosed by thick, white floss.
Nomenclature Species name comes language of the Carribbean Taino people, and means literally 'giant tree'. The specific epithet *speciosa* means 'showy', describing the tree in flower.
Uses Cloth can be made from fibres

obtained from the bark; cushions and mattresses can be stuffed with the floss (kapok) in seed capsules; the soft, light wood is used as a cork substitute.

The sheer height of these majestic trees with their spiny trunks make for an arresting sight.

PINK TRUMPET TREE

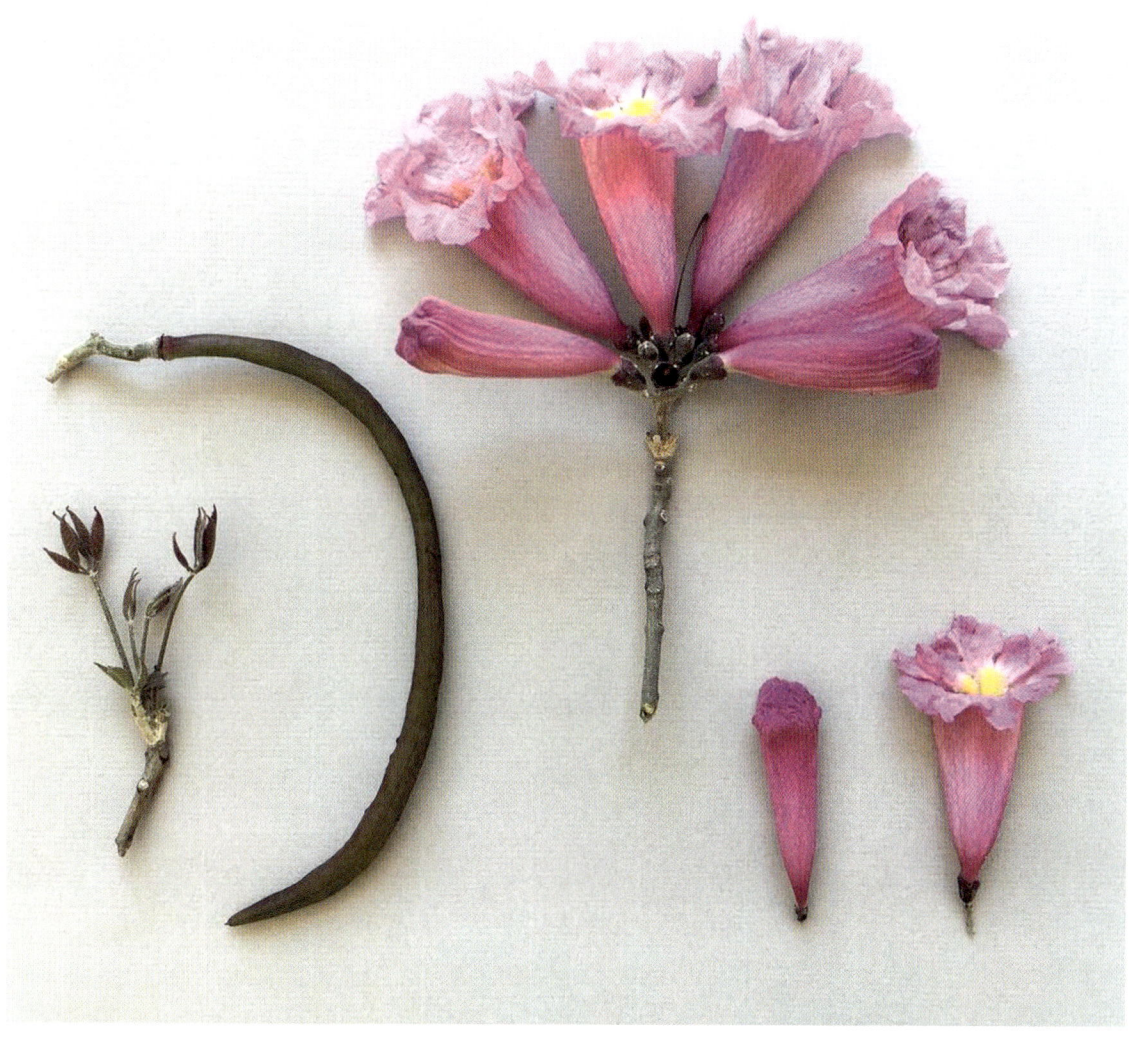

Latin name *Handroanthus impetiginosus* (syn. *Tabebuia palmeri*)

Photograph shows Young leaves, seed pod, inflorescence, flower, bud.

Collected Harare, Zimbabwe, 2015

Other common name Pink lapacho tree

Family *Bignoniaceae*

Origin Northern Mexico to South America.

Description A deciduous to semi-deciduous tree that grows up to 12m in height, commonly planted in gardens.

Leaves Palmately compound leaves are made up of five to seven leaflets with toothed margins, each up to 8cm long.

Flowers Borne in clusters, the showy, pink, tubular flowers with white throats are up to 6cm long. Usually flowers before the new leaves appear in the southern hemisphere or after leaf fall in northern hemisphere countries such as India.

Fruit Forms smooth, narrow, 30-cm long, green, dehiscent seed pods that turn brown as they ripen, then split to release winged seeds.

Nomenclature Species is named after the Brazilian botanist Oswaldo Handro (1908–86) and the Ancient Greek word *anthus* meaning 'flower'. *Impetiginosus* comes from the Latin *impeter* (to attack) referring to the skin condition impetigo that it was thought the plant could be used to treat; there is no evidence of this having been successful.

Uses Decorative ornamental tree. It attracts bees so is used a honey plant. The tree has multiple uses in traditional medicine, for example a tea made from the bark is an expectorant for upper respiratory infections.

This is the national tree of Paraguay. More familiar with the sulphur-yellow coloured flowers of *Handroanthus chrysanthus*, I was delighted to come across a single specimen of this tree in Harare's National Botanical Gardens.

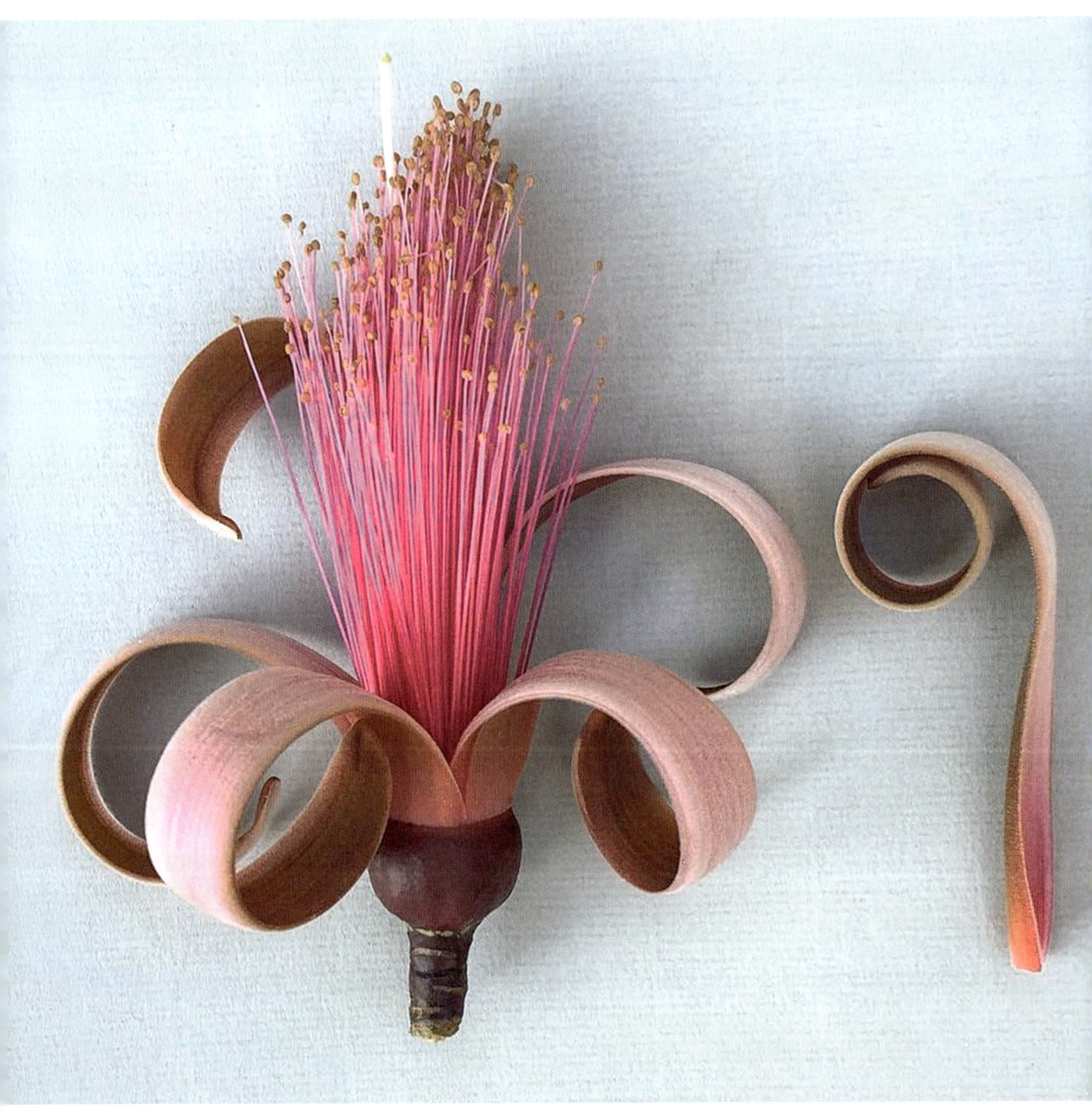

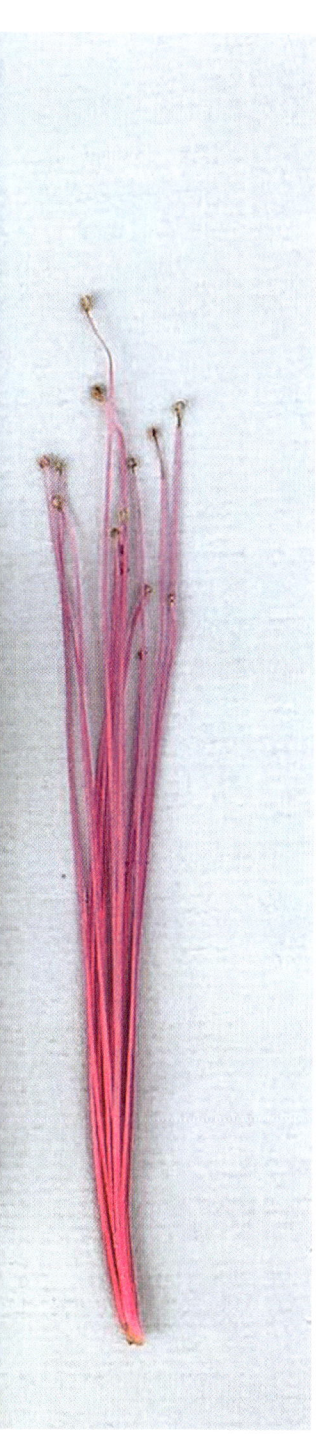

SHAVING BRUSH TREE

Latin name *Pseudobombax ellipticum*
Photograph shows Flower, sepal, stamens.
Collected National Botanical Gardens, Harare, Zimbabwe, 2016
Other common name Ampolla tree
Family *Malvaceae* (formerly *Bombacaceae*)
Origin Southern Mexico, El Salvador, Guatemala, Honduras.
Description A fast-growing, deciduous tree that is often around 10m high, but can be as much as 15–21m tall, and has a spread of 1.3m. It has a swollen, smooth, green trunk and fissured grey bark.
Leaves Palmate leaves have five elliptical to rounded, dark-green leaflets, each up to 30cm long and 18cm wide.
Flowers Upright, black buds appear in spring, when the tree is devoid of leaves. As they open, tan-coloured sepals peel back and curl downwards to reveal beautiful pink 'tassels' comprising 'bunches' of multiple 12-cm stamens. The flowers are largely pollinated by bats and moths.
Fruit Forms elongated dehiscent capsules up to 15cm long that contain numerous seeds surrounded by silken white strands (kapok). Seeds are wind dispersed when capsules split open.
Nomenclature *Pseudobombax* is a combination of the Latin and Greek words *bombax*, meaning 'cotton', and *pseudo* or 'false' because the plant was previously classed in the genus Bombax. The epithet *ellipticum* describes the shape of the leaflets.
Uses Kapok from the seed capsules can be used to stuff pillows and as insulation. The seeds can be toasted and eaten. In El Salvador, parts of the tree are used traditional medicines – a tea made from the flowers is said to relieve upset stomachs and one made using the fresh bark is a treatment for diabetes.

My father has long been my botanical wingman, willing to pull over if I yell 'stop the car' when I see something of interest blooming in the veldt. He is unfailingly patient while I then set about collecting and/or photographing a specimen. Occasionally, also he sends me a photograph of a tree or plant to identify. Coming across a tree in Harare's Botanical Gardens covered in unusual pink flowers, he took a photograph and sent it to me for identification. A couple of years later when I was back in Harare at the same time of year, we returned to the Botanical Gardens and the tree was in bloom again. With permission I collected some flowers and captured this deconstruction.

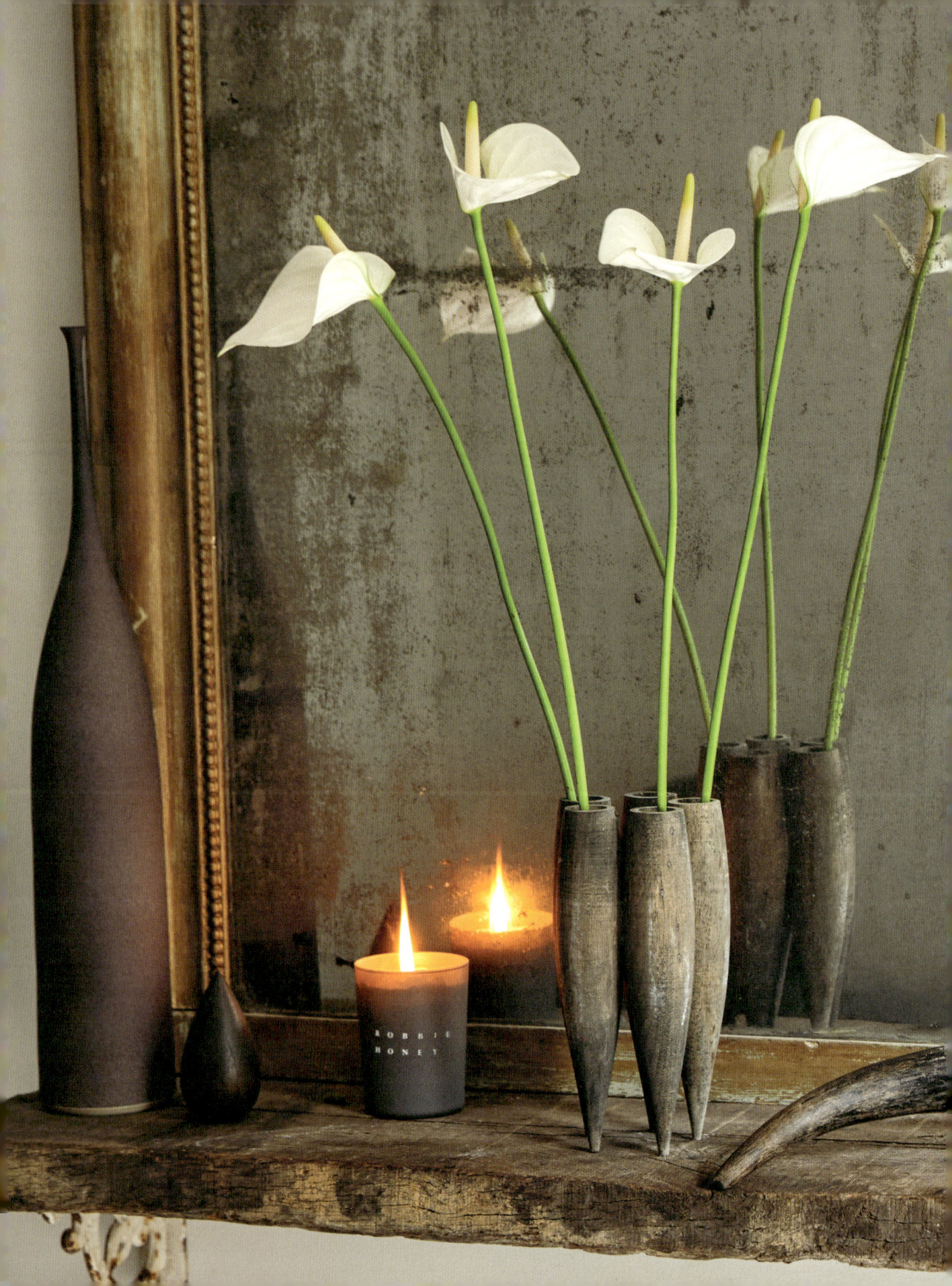

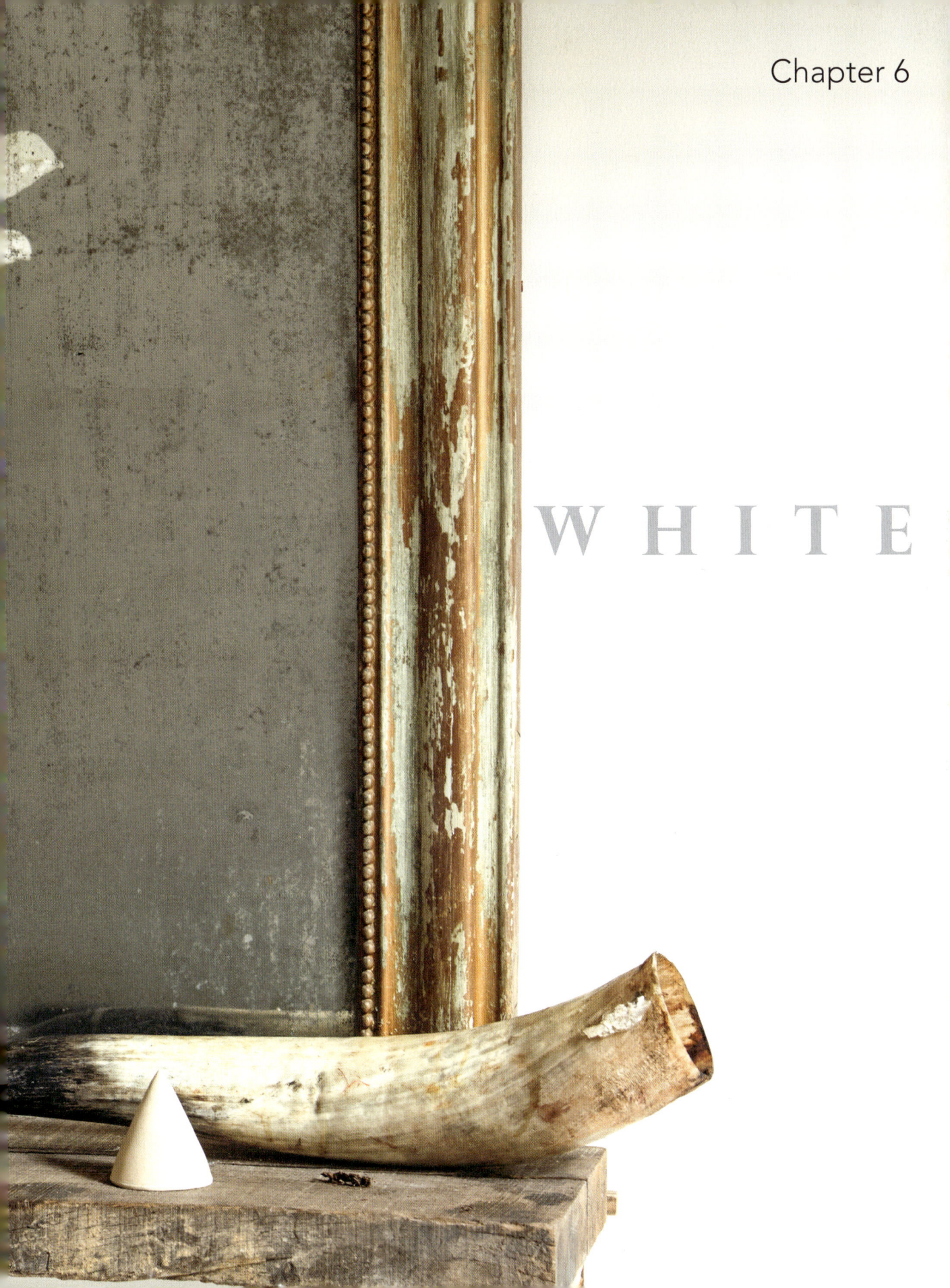

Chapter 6

WHITE

DAFFODIL MOUNT HOOD

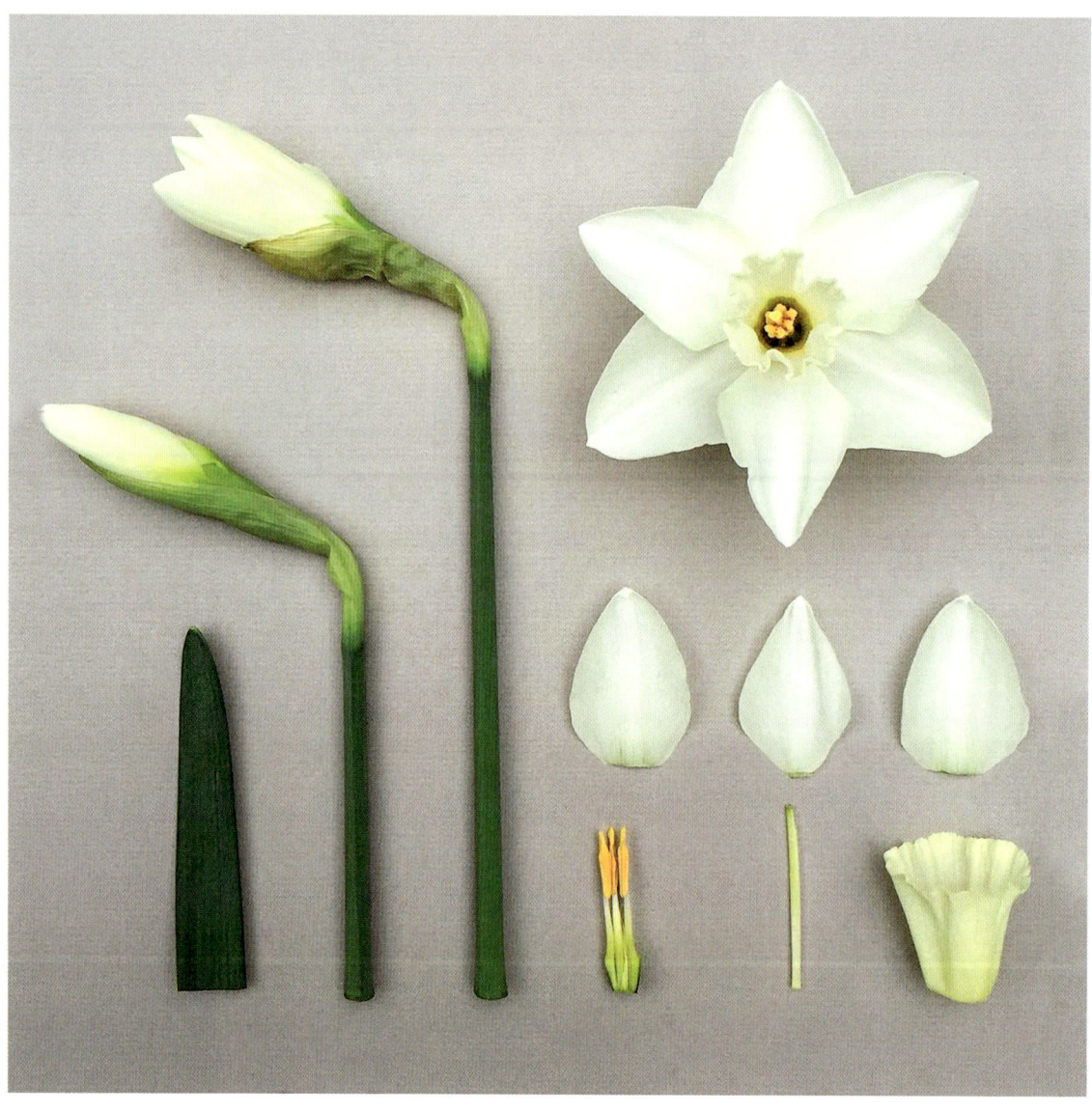

Latin name *Narcissus* 'Mount hood'
Photograph shows Leaf, buds (closed and partially open), flower, petals, corona, pistil, stamens.
Collected London, UK, 2015
Other common name Trumpet daffodil
Family *Amaryllidaceae*
Origin Holland
Description A robust, spring-flowering, bulbous perennial cultivar that can reach a height of 45cm. *Narcissus* 'Mount hood' was bred in Holland in the 1930s.
Leaves Light-green, strap-like, erect leaves are covered in a colourless sheath.
Flowers Bears a long-lasting, single, ivory-white flower, up to 10cm across, on a leafless stem. Each flower has six petals.

Fruit Not ornamentally significant – it typically forms three-sectioned seed pods, 15–20mm in length.
Nomenclature *Narcissus* comes from Latin *narcoticus* and Greek *nàrkissos* derived from *narke* meaning numbness, or torpor – the effect of a narcotic. In classical mythology Narkissos (or Narcissus) was a beautiful youth who fell in love with an image of a man he observed in a pool, not realising that it was own reflection. Transfixed by the beauty of the image he saw Narkissos could not tear his gaze away from it, so remained staring at his own reflection until his death.
Uses Decorative ornamental.

Growing up in Zimbabwe, daffodils were rare. I tried to grow them, but only succeeded in getting one bulb to flower in the shade of an avocado tree. At the age of 17, I arrived in the UK to study. In February, I found every verge awash with yellow daffodils and by the end of March my holy grail had entirely lost its appeal. Today my perfect daffodil is a white one.

TRIANDRUS DAFFODIL

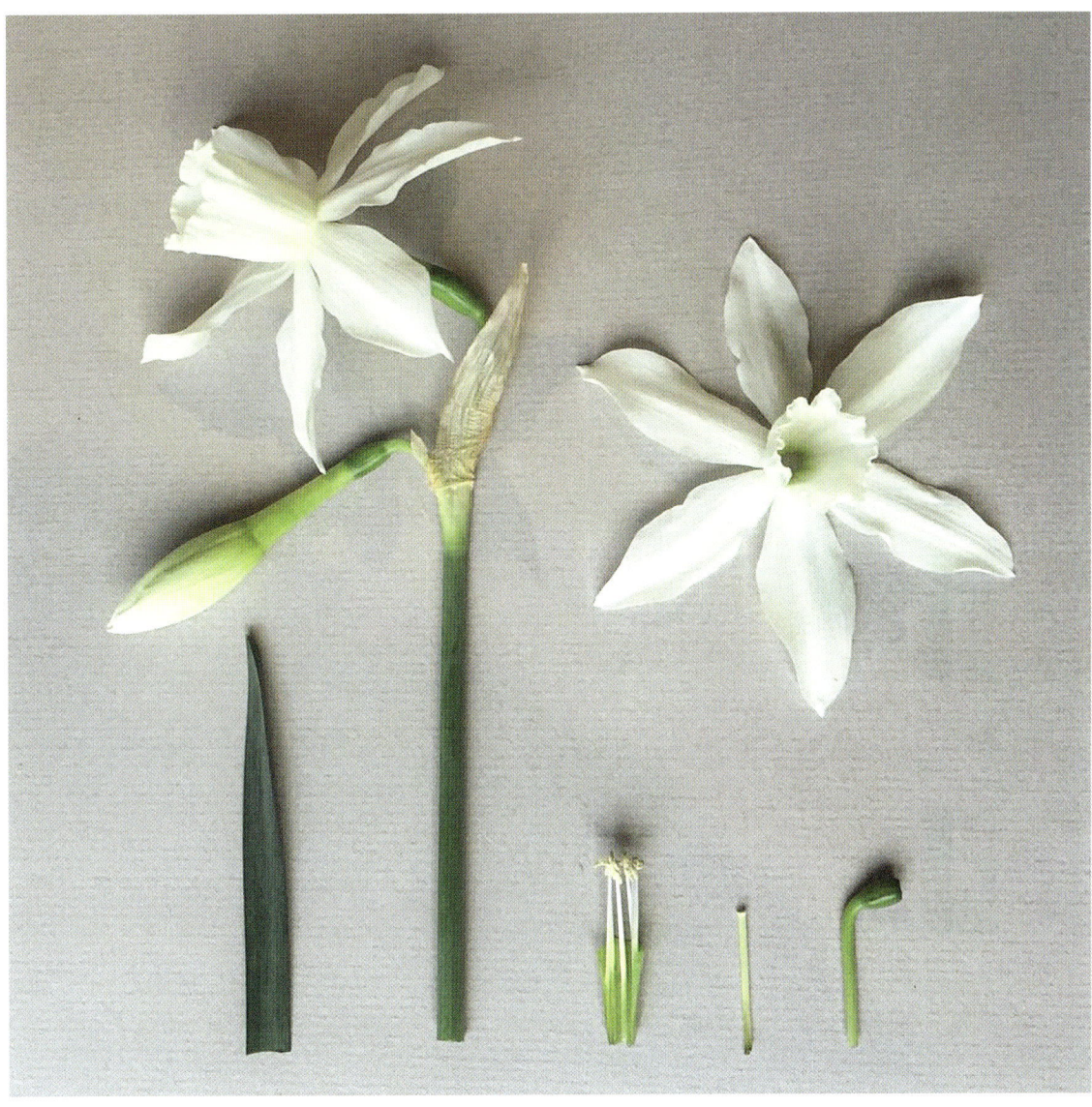

Latin name *Narcissus* 'Thalia'
Photograph shows Leaf, flower and bud, flower, neck of flower, pistil, stamens.
Collected London, UK, 2014
Family *Amaryllidaceae*
Origin Holland
Description A hardy, award-winning, spring-flowering, bulbous perennial daffodil cultivar developed in 1916 that grows to 20cm in height.
Leaves Forms deep-green, narrow, strap-like leaves.
Flowers The whitest of all *narcissus* species, the fragrant, vigorous, bell-shaped flowers with flexed petals of *N.* 'Thalia' are borne on leafless stems that each carry up to four blooms.
Fruit Long-lived bulbs grow up to 16cm.
Nomenclature *Narcissus* comes from Latin *narcoticus* and Greek *nàrkissos* derived from *narke*, meaning numbness, or torpor – the effect of a narcotic. In classical mythology Narkissos was a beautiful youth who fell in love with his own image (see opposite).
Uses Decorative ornamental.

Friends had recently given birth to a baby who I heard, by word of mouth, had been named Thalia. By way of congratulations I ordered some bulbs of *Narcissus* 'Thalia' for them, but I then discovered that the baby's name was in fact Talia – without the 'H'. I gave them some of the bulbs anyway, but I also kept some for myself, and the flowers I photographed here are the result of my mistake. Captured in spring sunshine, these narcissi light up a garden.

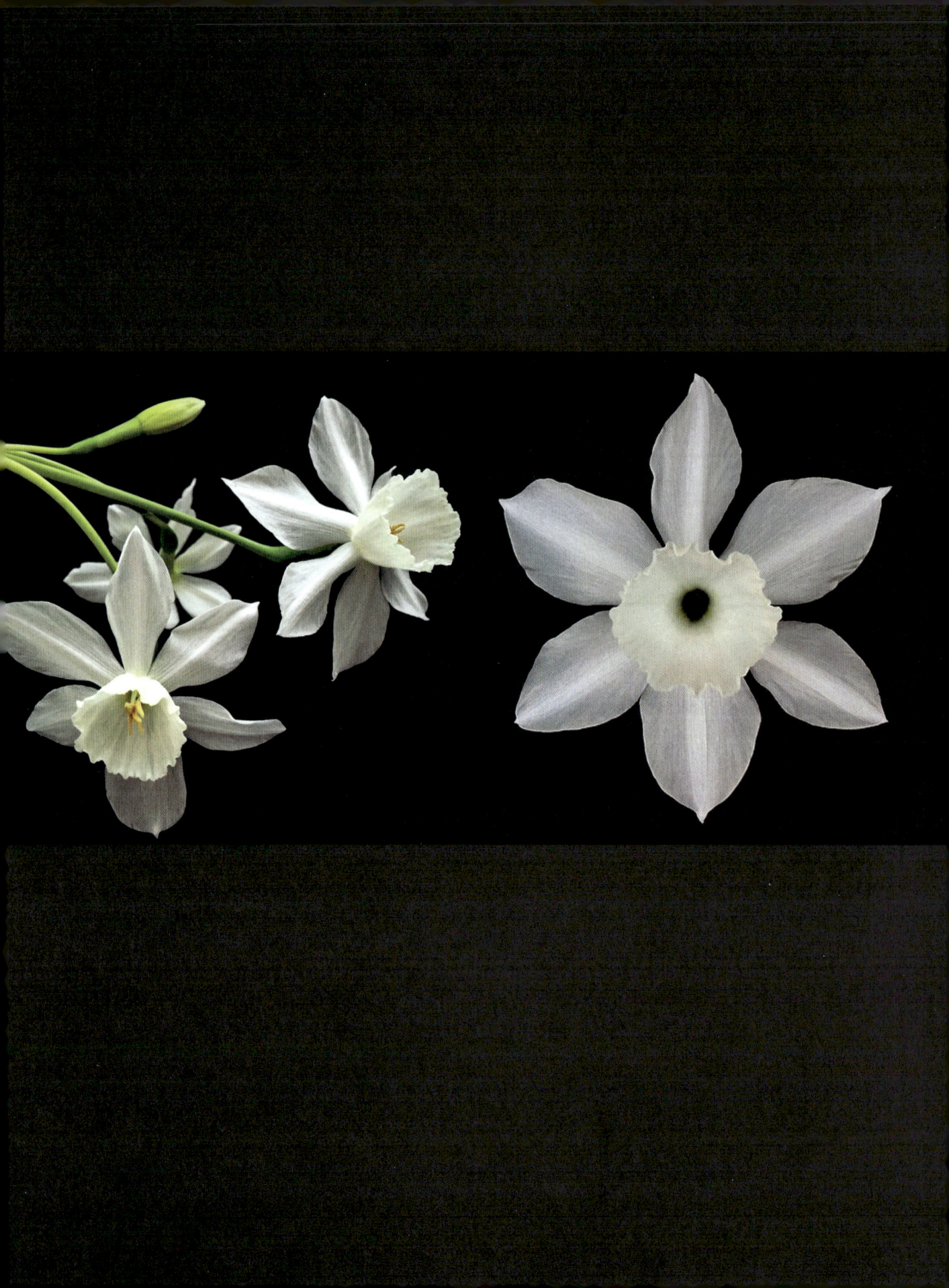

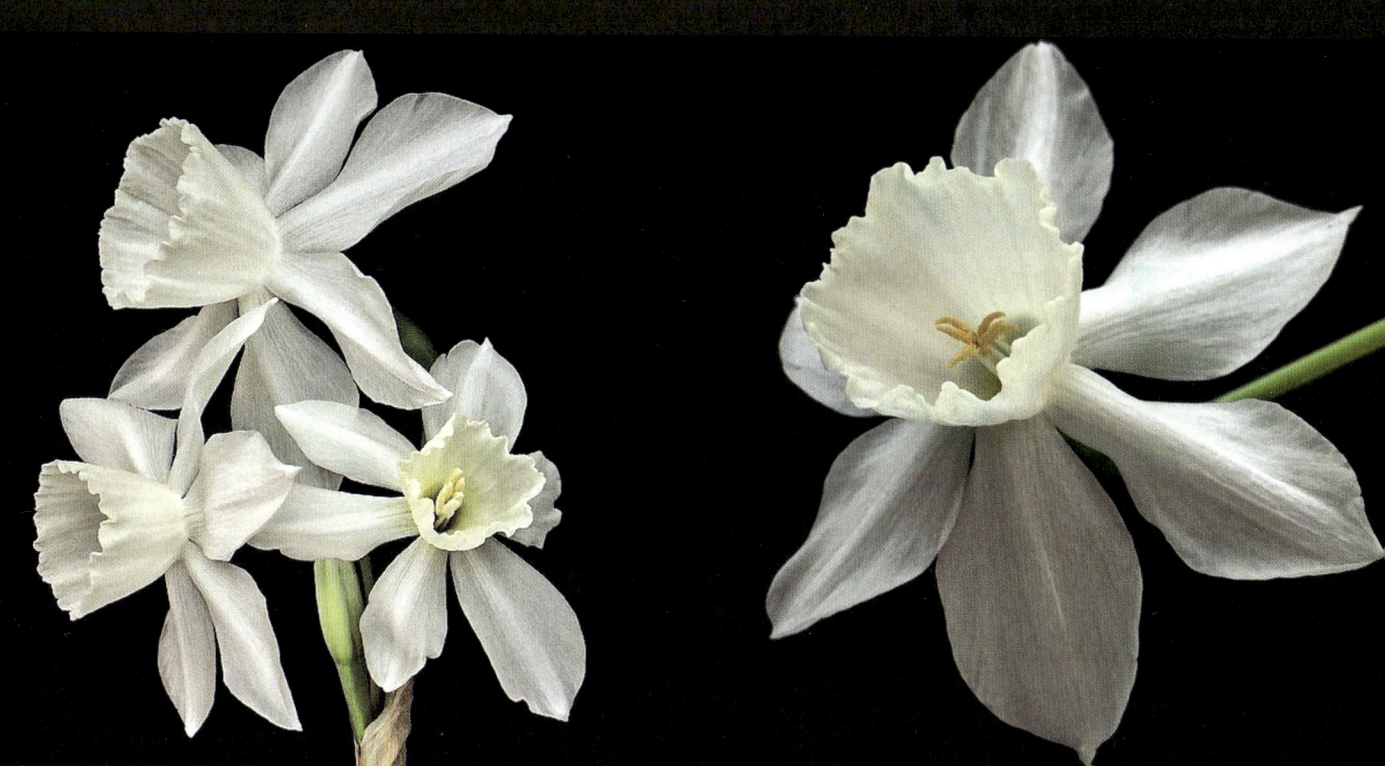

Along with the Poet's daffodil *Narcissus poeticus*,
Narcissus 'Thalia' is one of my favourite Narcissi.
The flowers with their reflexed petals are almost
luminescent. I grew these specimens from bulbs I
purchased in Beijing, China. Their flower shapes
differ slightly from other *N.* 'Thalia' I have grown.
As can be seen from these photographs, the stems
bear as many as four flowers.

DUTCH IRIS

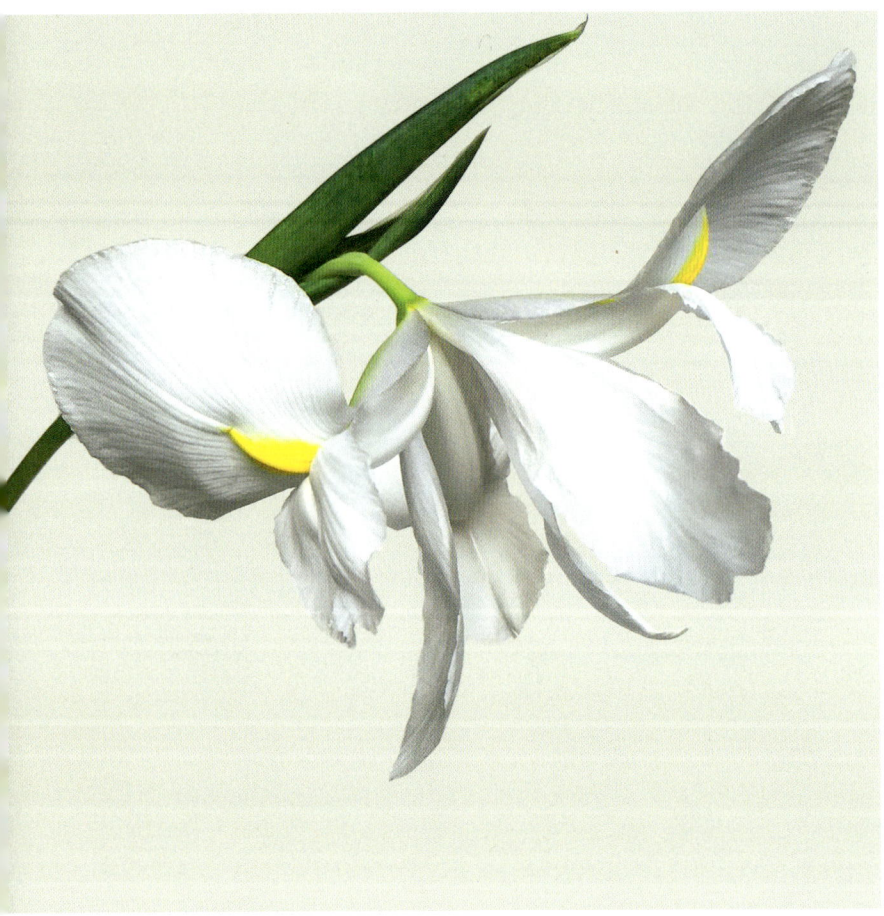

Latin name *Iris* x *hollandica*
Photographs show Left: flower head on stem. Opposite: fall petal, flower head, standard petals, stamens.
Collected London, UK, 2014
Family *Iridaceae*
Origin A hybrid grown from *Iris tingitana* and *Iris xiphium*. Parent plants are native to North Africa (*I. tingitana*) and Spain (*I. xiphium* var. *praecox*).
Description A summer-flowering perennial hybrid grown from a bulb that can be 60cm tall.
Leaves Forms slender, grass-like leaves.
Flowers Bears large white, or blue to yellow to flowers that are 10cm in diameter.
Fruit The fruits are capsules.
Nomenclature Species is named after Iris, the Greek goddess of the rainbow and messenger of the gods. *Hollandica* refers to the fact that this hybrid was created in Holland.
Uses Decorative ornamental.

This iris species' elegant crystalline white petals splashed with yellow, really stand out when their green casing is stripped away.

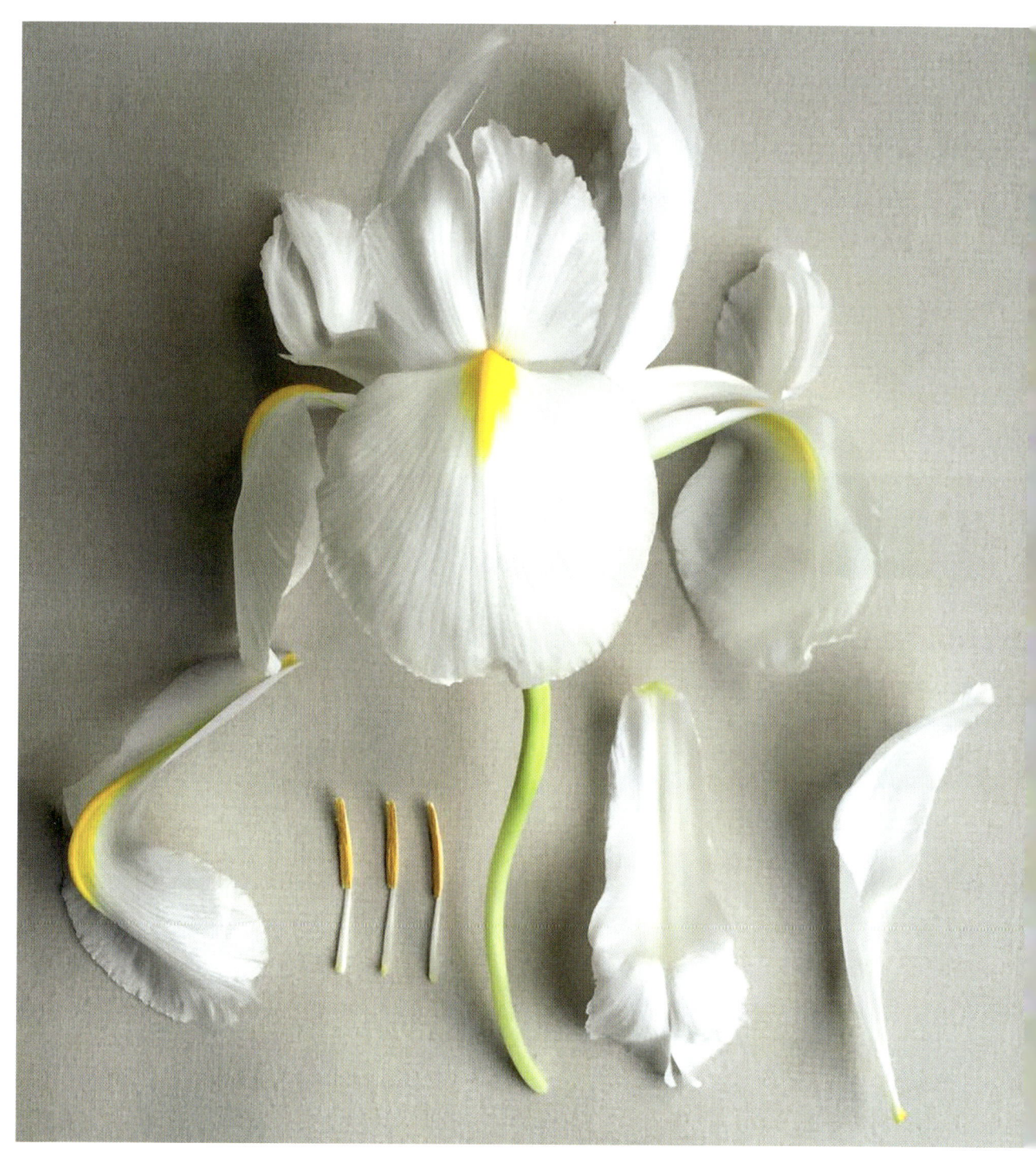

VLEI LILY

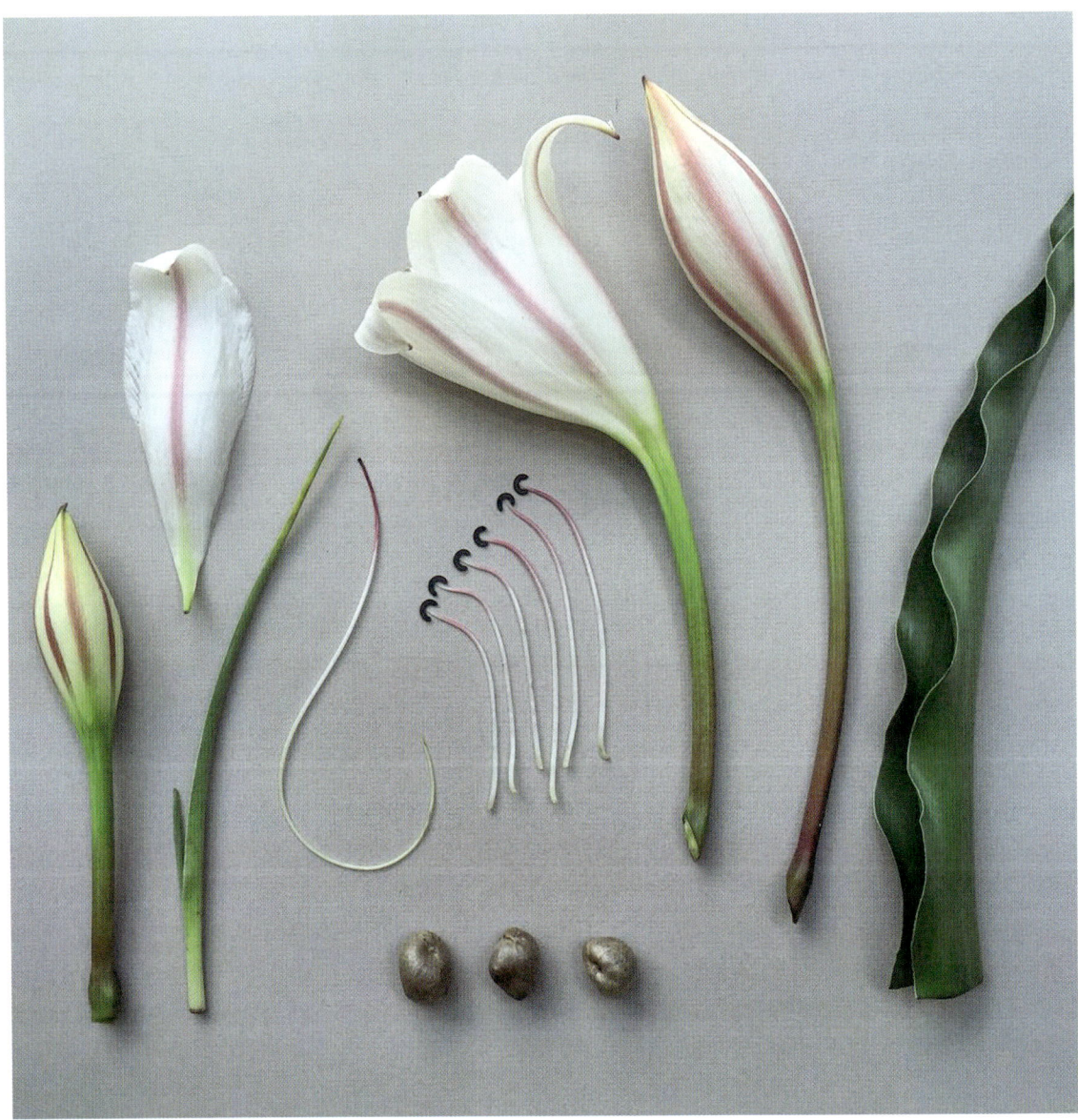

Latin name *Crinum macowanii*
Photographs show Above: bud, petal, flower, mature bud, leaf, seeds, stamens, pistil, stem with bud. Opposite: stamens.
Collected Komani airstrip, Zimbabwe, 2015
Common name River crinum lily, Cape coast lily, River lily.
Family *Amaryllidaceae*
Origin Southern Ethiopia, Zambia, Malawi, Botswana, Zimbabwe.
Description Grown from a bulb, this

deciduous, summer-flowering, plant can grow to a height 1.1m.
Leaves Strap-like, large, green, fleshy, variable leaves, grow up to 1m long and 2–20cm wide. They have a slight bluish tint and undulating margins.
Flowers Large, bell-shaped, fragrant, pink and white flowers are borne on long stalks and produced in umbels of up to 25 blooms.
Fruit Forms capsules containing up to six irregularly shaped, fleshy, green

seeds with a water-repellent covering.
Nomenclature The genus name *Crinum*, comes from the Greek word *krinon*, meaning lily. This particular species is named after British botanist and teacher, Peter MacOwan (1830–1909), who collected extensively in southern Africa.
Uses Widely used in traditional medicine in southern Africa. The bulbs stimulate the milk production and treat kidney and bladder infections. The leaves can be utilised as bandages or as a poultice.

After the first rains, Harare's Vleis, or wetlands, erupt with an abundance of white mushrooms. When this happened we used to go *en famille* on 'mushroom walks', armed with baskets to collect them. From an early age, we were taught to discern which ones were safe to eat and which were not. Our test was to attempt to peel the caps: if the skin could be peeled, it could be eaten, and if not, it was to be avoided. We would often eat them straight from the ground, savouring their earthy freshness. For me, however, the highlight of these walks was not the fungi, but the Vlei lilies

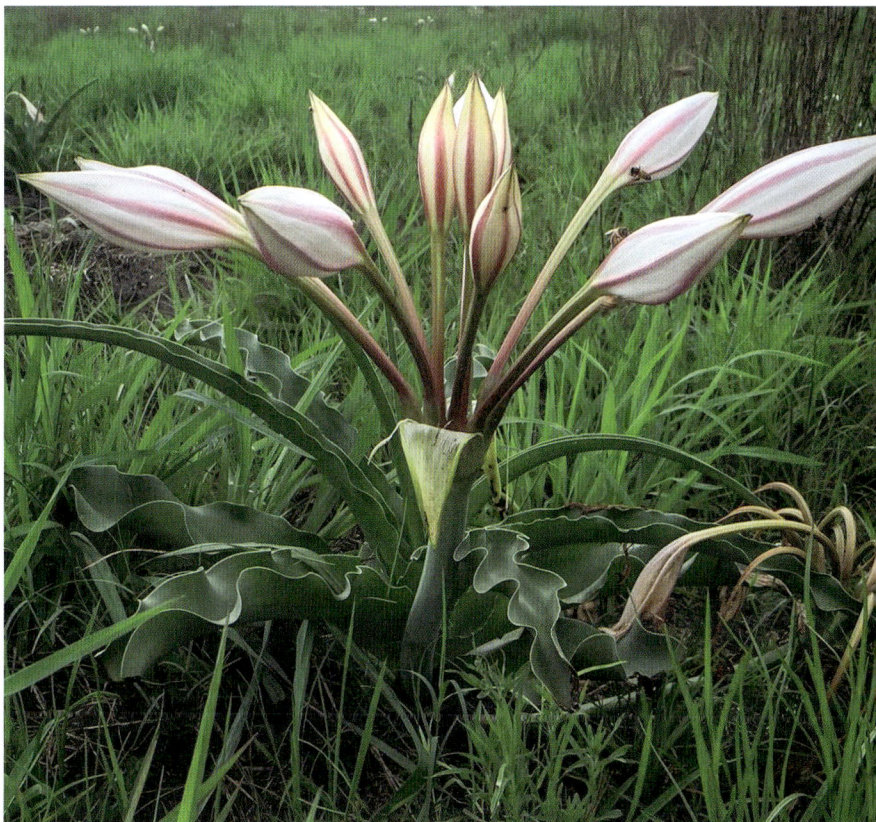

that also bloomed after the rains. I collected this specimen on the edge of the Komani airstrip after a joyride with my brother in his plane. The candy-striped buds of this *amaryllid* open at dusk, but the flowers also spill their delicious scent from unopened buds. Bees gather in a frenzy of expectation, on one occasion I observed them wriggling their way into the unopened buds 'burgling' the flowers of their nectar and pollen before nightfall. Aside from bees, I understand that these, like many other night-scented flowers, are mostly pollinated by hawk moths.

TAMANU

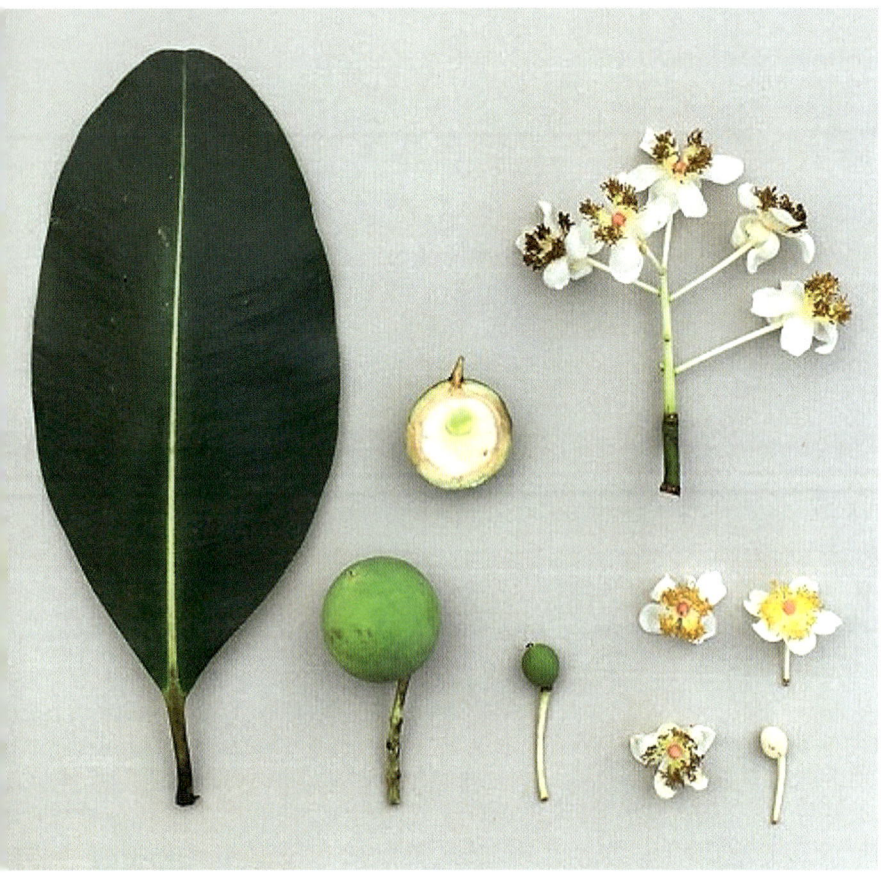

I first encountered this tree on a beach on an island on Malaysia's Langkawi Archipelago and was struck by the scent of its pretty flowers.

Latin name *Callophylum innophylum*
Photograph shows Mature leaf, section of fruit showing seed, inflorescence, flowers, bud, young fruit, mature fruit.
Other common names Due to its worldwide distribution, it has many common names, including Indian Laurel, Alexandrian laurel, and Ball tree.
Family *Clusiaceae*
Origin Sandy beaches and coastal forests of East Africa, Asia from southern coastal India and Sri Lanka to Japan, Vietnam, Thailand, Taiwan, Philippines, Fiji, Malaysia, Singapore, Papua New Guinea, and Australia.
Description A large, evergreen, flowering tree that grows to a height and spread of 25–30m. It has long, spreading branches with dark, fissured bark, and a broad, uneven crown.
Leaves Glossy, dark-green, elliptic, stiff, leathery leaves with entire margins grow to 4–8cm long.

Flowers Bears inflorescences with profusions of up to 15 small, sweet-scented, white flowers, arranged in panicles. Blooms have cupped petals and fluffy yellow stamens and only last one day.
Fruit Forms smooth, round drupes – up to 4cm in diameter – that contain one large seed. Dull-green at first, the colour varies from yellow to red when ripe, and the fruit becomes wrinkled.
Nomenclature *Calophyllum* translates into 'beautiful leaf', from the Greek words *kalos* meaning 'beautiful', and *phullon* or 'leaf'.
Uses It has many uses. In particular, the wood is used in construction and boat-building and the dark-brown, dried seeds, or nuts, yield a rich, dark-green, fragrant oil (fresh seeds do not contain any oil). 'Tamanu oil' is used extensively in skincare products and is fêted for its regenerative and healing properties.

SNOW WHITE SLIPPER ORCHID

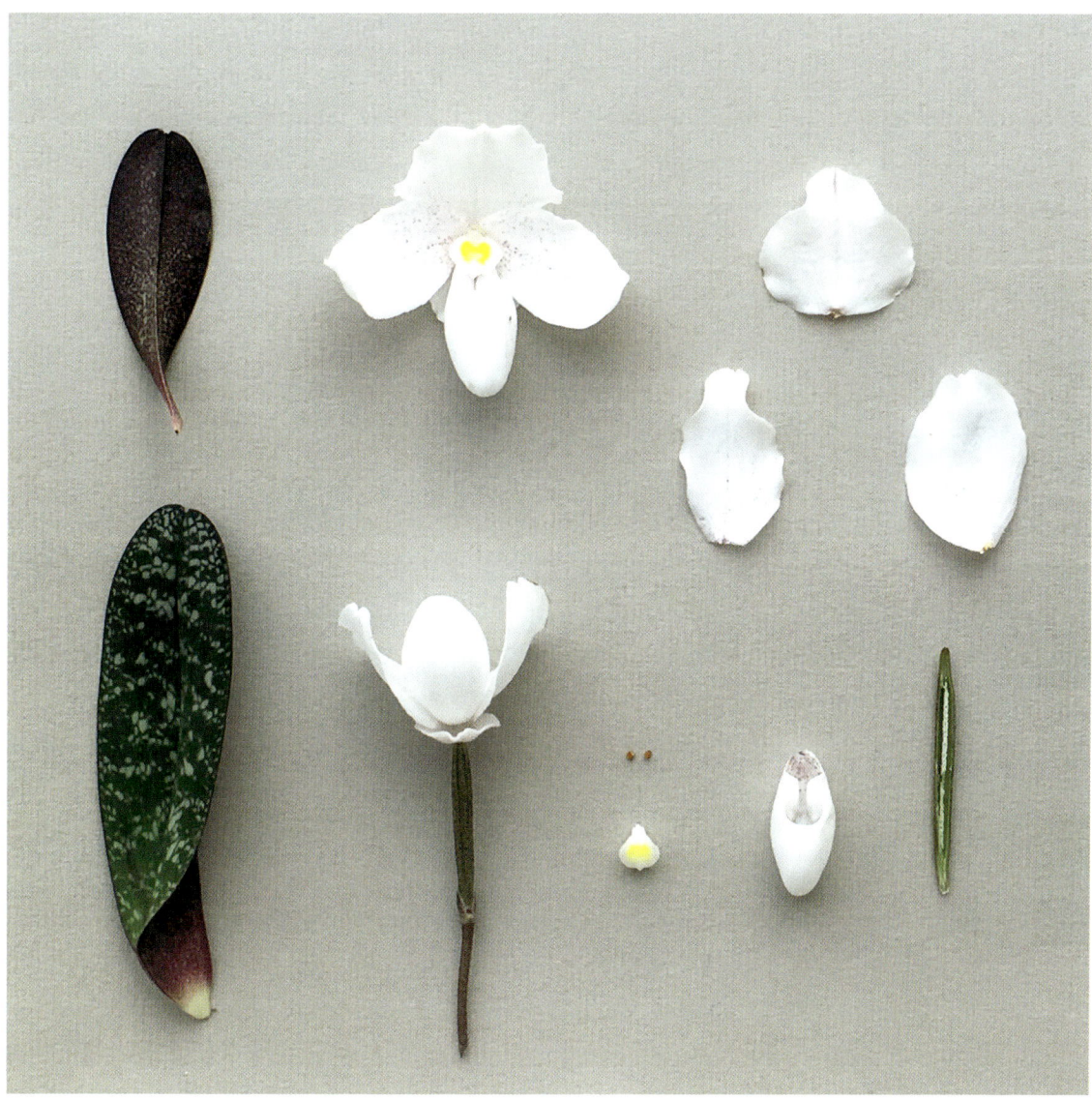

Latin name *Paphiopedilum niveum*
Photograph shows Mature leaf, young leaf, flower, petals, seed capsule, pouch, staminode, pollen, inverted flower on a stem.
Collected Langkawi, Malaysia, 2017
Other common name Its Thai name is *Rongthao nari dok khao satun.*
Family *Orchidaceae*
Origin Archipelago of Palau Langkawi, Malaysia; Thailand; Myanmar; Borneo and surrounding islands.
Description A diminutive, lithophytic orchid that grows to 10cm in diameter.
Leaves Dark-green, strap-like leaves are marbled silver grey and have purple, speckled undersides.

Flowers The small, crystalline white flowers, delicately flecked with varying degrees of purple, measure 5–6cm in diameter. They have elliptic petals, a small lip, an ovoid pouch, and a bright yellow staminode. Less commonly seen is the pure white form *P. niveum* 'Album'.
Fruit Produces a 2cm-long seed capsule.
Nomenclature *Paphiopedilum* comes from the Greek words *paphos* referring to the temple of Aphrodite, and *pedilon* meaning 'shoe'. The epithet *niveum* is from the Latin *niveus*, meaning 'snowy', referencing the white blooms.
Uses This species has been used to parent many white hybrid *Paphiopedila* orchid plants.

This orchid is increasingly rare and now listed as an endangered species due to ruthless collecting and loss of habitat. I was lucky enough to be taken to a secret location on the Malaysian peninsula by Irshad Mubarak, Langkawi's naturalist). Climbing the limestone cliffs above the ocean to see these orchids and collect some flowers was a high point of my year's botanizing.

MAGNIFICENT AERANGIS

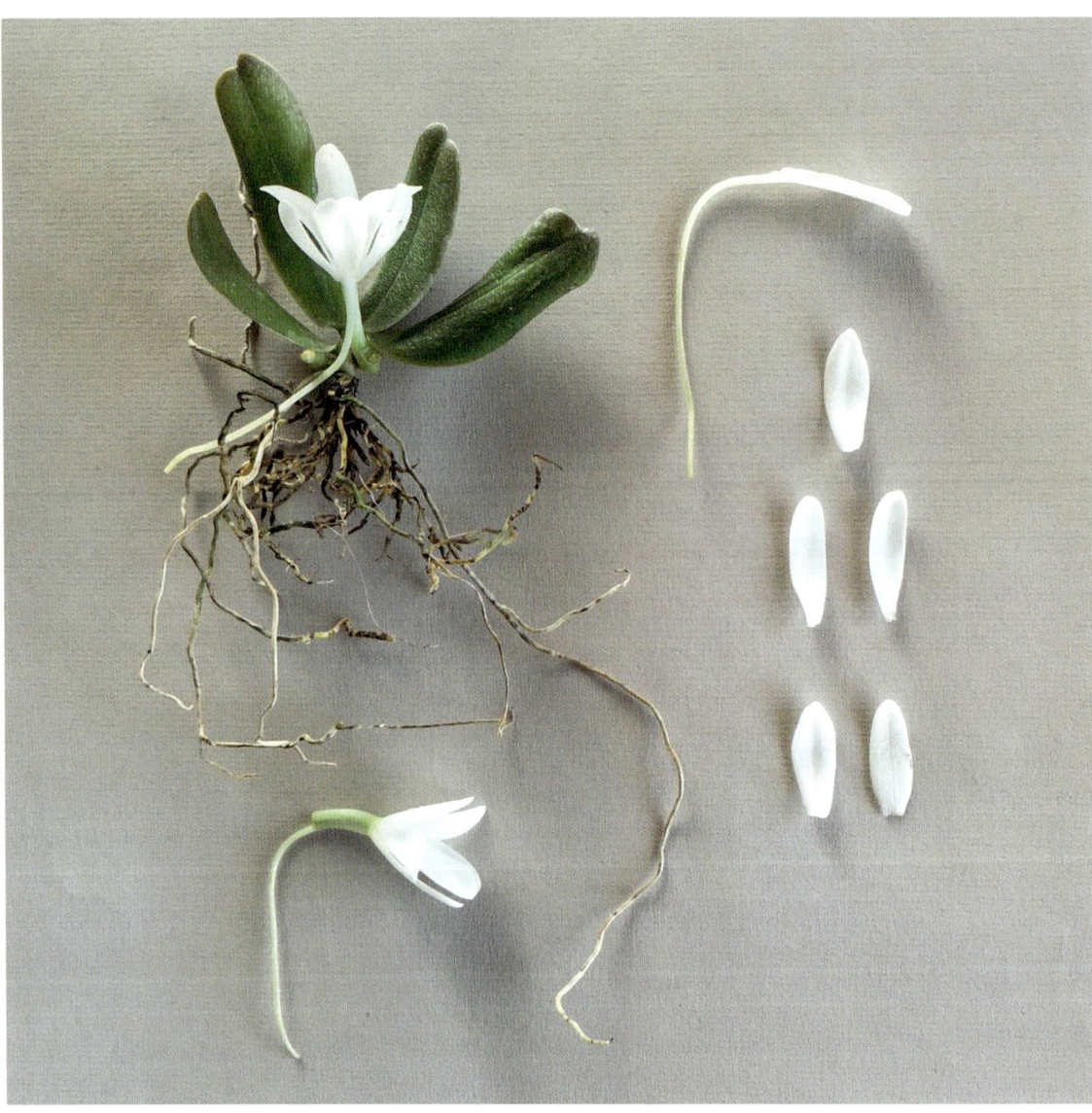

Latin name *Aerangis fastuosa*
Photograph shows Plant with leaves and flowers, nectary and lip, petals and sepals, flower.
Collected Harare, Zimbabwe, 2016
Family *Orchidaceae*
Origin Between coastal plains and central plateau of Madagascar.
Description A dwarf, hot-growing, epiphytic orchid found on twigs and branches in evergreen forests at elevations between 1,000–1,500m.
Leaves These grow up to 10cm and are borne on a short stem about 3cm long.
Flowers Large, long-lasting, night-scented, white flowers grow up to 5cm in diameter; each spike can bear from one to six flowers.
Fruit None
Nomenclature Genus name *Aerangis* is from the Greek words, *aer*, meaning 'air' and *angos*, or 'vessel', in reference to the spur at the base of the lip. *Fastuosa* is Latin for 'proud'.
Uses Decorative ornamental

Indigenous *aerangis* species orchids were the first Zimbabwean native examples I grew. I bought this tiny *Aerangis fastuosa* from Zimbabwean plantsman John Hibbert, and was able to capture its diminutive flowers with their splendidly long nectaries – slender, nectar-filled tubes situated behind the flower. Deliciously scented, this orchid is pollinated by hawk moths whose proboscides are the same length as the flower's nectaries.

ANGEL ORCHID

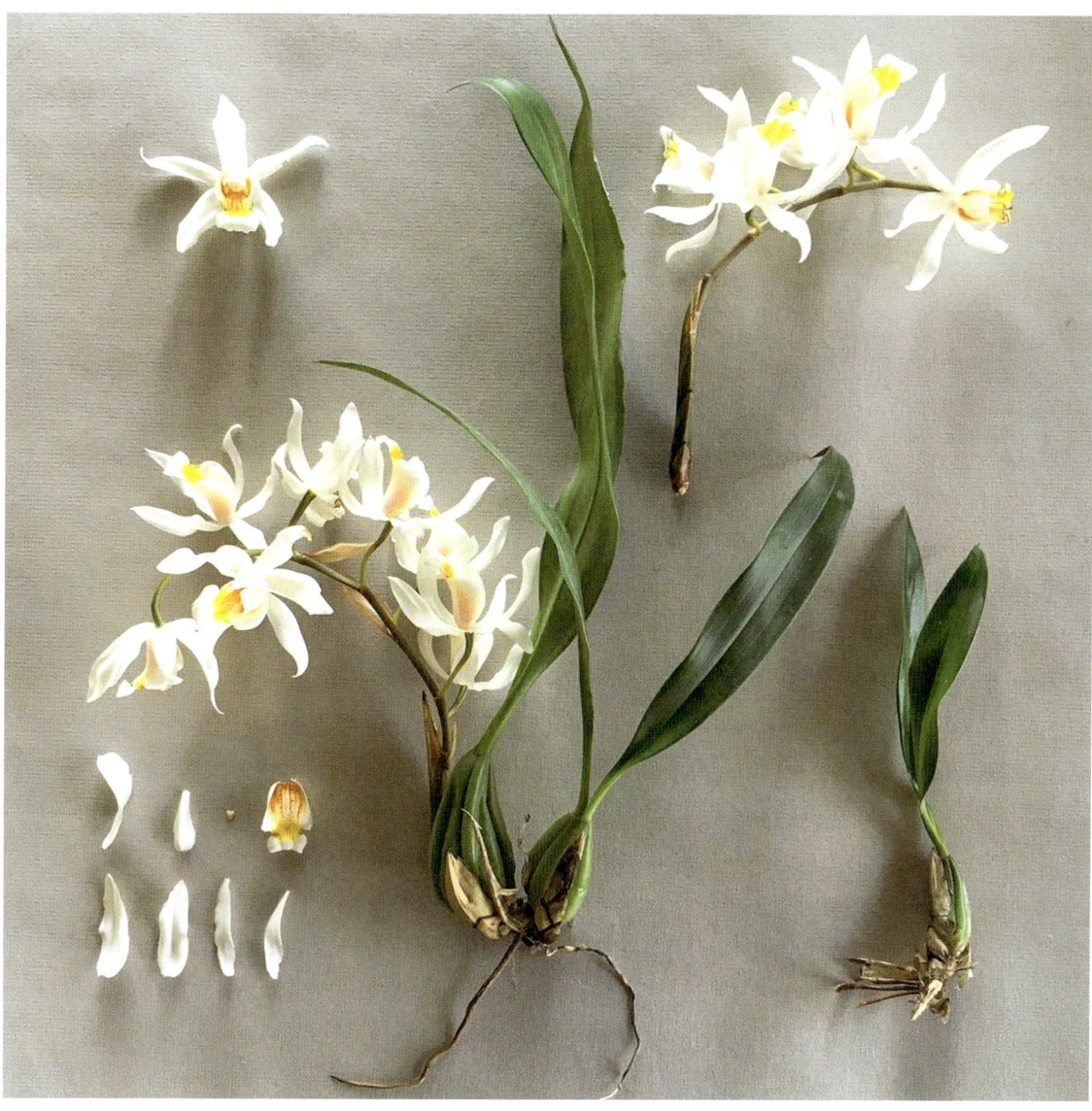

Latin name *Coelogyne cristata*
Photograph shows Flower, pseudo-bulbs with leaves and flowers, detached flower spike, pseudo-bulb, spike sepals, petals, labellum, column, anther cap, pollinia.
Collected Melfort, Zimbabwe, 2017
Other common name Snow queen
Family *Orchidaceae*
Origin From lowland forest to high altitudes of India, southwest China, Philippines, Indonesia, and New Guinea
Description A sympodial, epiphytic and lithophytic orchid, with rounded pseudo-bulbs.

Leaves Lance-shaped, narrow, pointed leaves are usually bright green, 15–30cm long, and 5cm wide.
Flowers Scented, white flowers (5–6cm) with yellow fuzzy marking on lip and ruffled petals, grow on a long (30cm) stalks bloom in winter and spring.
Fruit None
Nomenclature *Coelogyne* is from the Greek words *koilos*, meaning 'hollow', and *gyne*, or 'woman', referring to the deep-set stigmatic cavity found in members of the genus. The epithet *cristata* means 'having a comb or tuft'.
Uses Decorative ornamental.

A friend, Tana and her family live just outside Harare in a thatched farmhouse set amongst acacia trees. When I visit I often see a pair of giraffe grazing the trees in front of the house. Tana and I collected this specimen from her neighbour's shade house. The scent of its flowers are reminiscent of horse manure, but in a good way!

ANEMONE

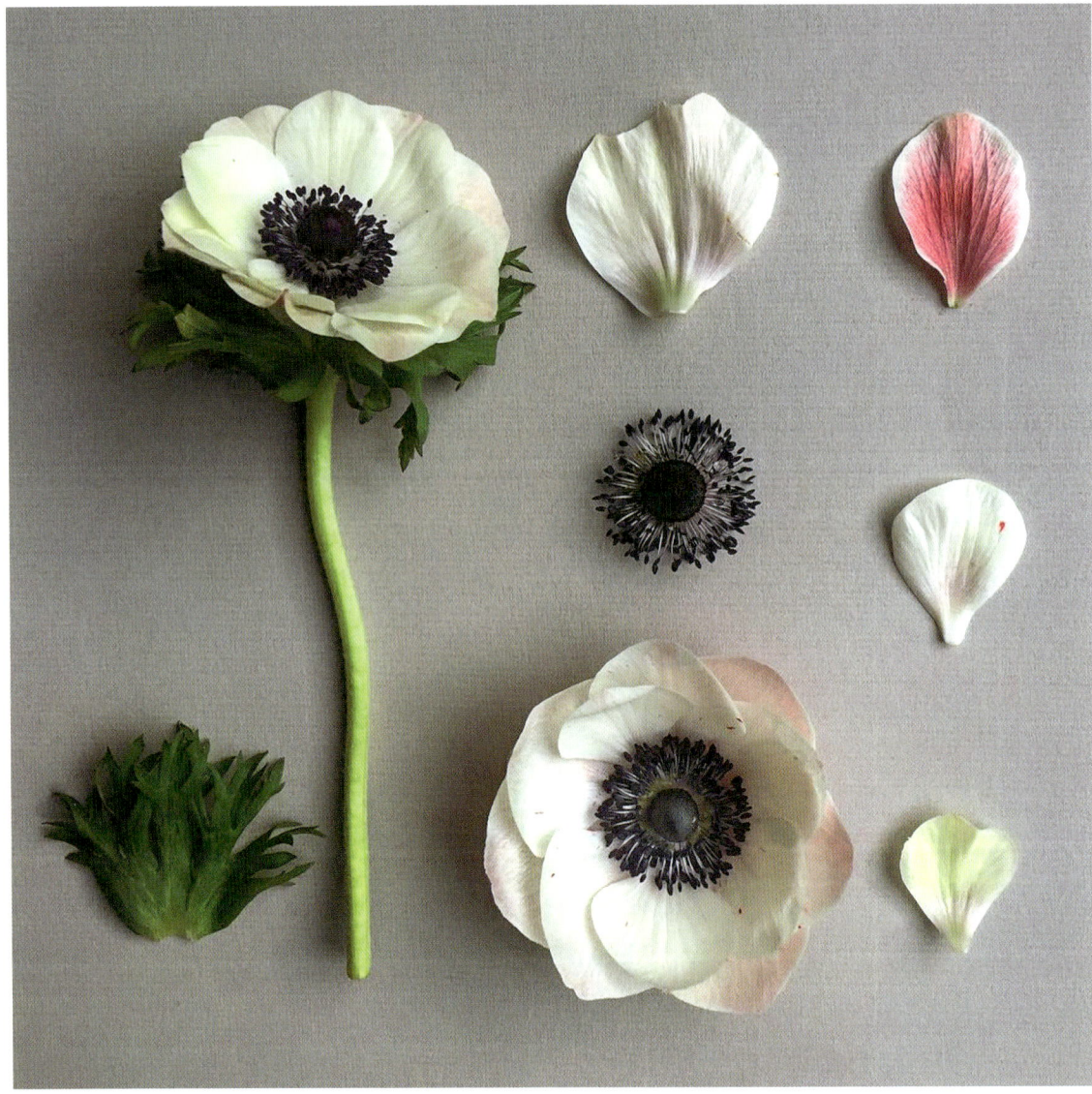

Latin name *Anemone coronaria*
Photograph shows Leaf, stalk and flower, petals, sepals, flower, corolla.
Collected Seoul, South Korea, 2016
Other common name Windflower
Family *Ranunculaceae*
Origin Mediterranean
Description Herbaceous, spring-flowering, tuberous perennial that can reach a height and spread of 50cm.
Leaves Basal leaves grow in a rosette formation with long stems and have three deeply lobed leaflets. Leaf margins can be dentate or entire.
Flowers Solitary, cup-shaped white, red, or violet flowers bloom in spring and can be up to 7cm in diameter. Each one has five petals and a ring of stamens (corolla). After flowering, the plant becomes dormant.
Fruit Produces small achenes with plumose tails. Hard, black, tubers also form, which function as storage organs and can be used to propagate plants.
Nomenclature *Anemone* is from the Greek *anemos* meaning 'windflower'. In Greek mythology, anemones sprang from the Goddess Aphrodite's tears as she mourned the death of Adonis. *Coronaria* means 'crown' and describes flower's central stamens.
Uses Decorative ornamental.

On a trip to Seoul one spring, the cherry trees were in bloom and I used huge bundles of their flowers to decorate a room. At the bases of my 'blossom trees' I amassed hundreds of these pretty anemones flushed palest blush pink – they were a perfect foil for the pink and white cherry blossom.

CHRISTMAS ROSE

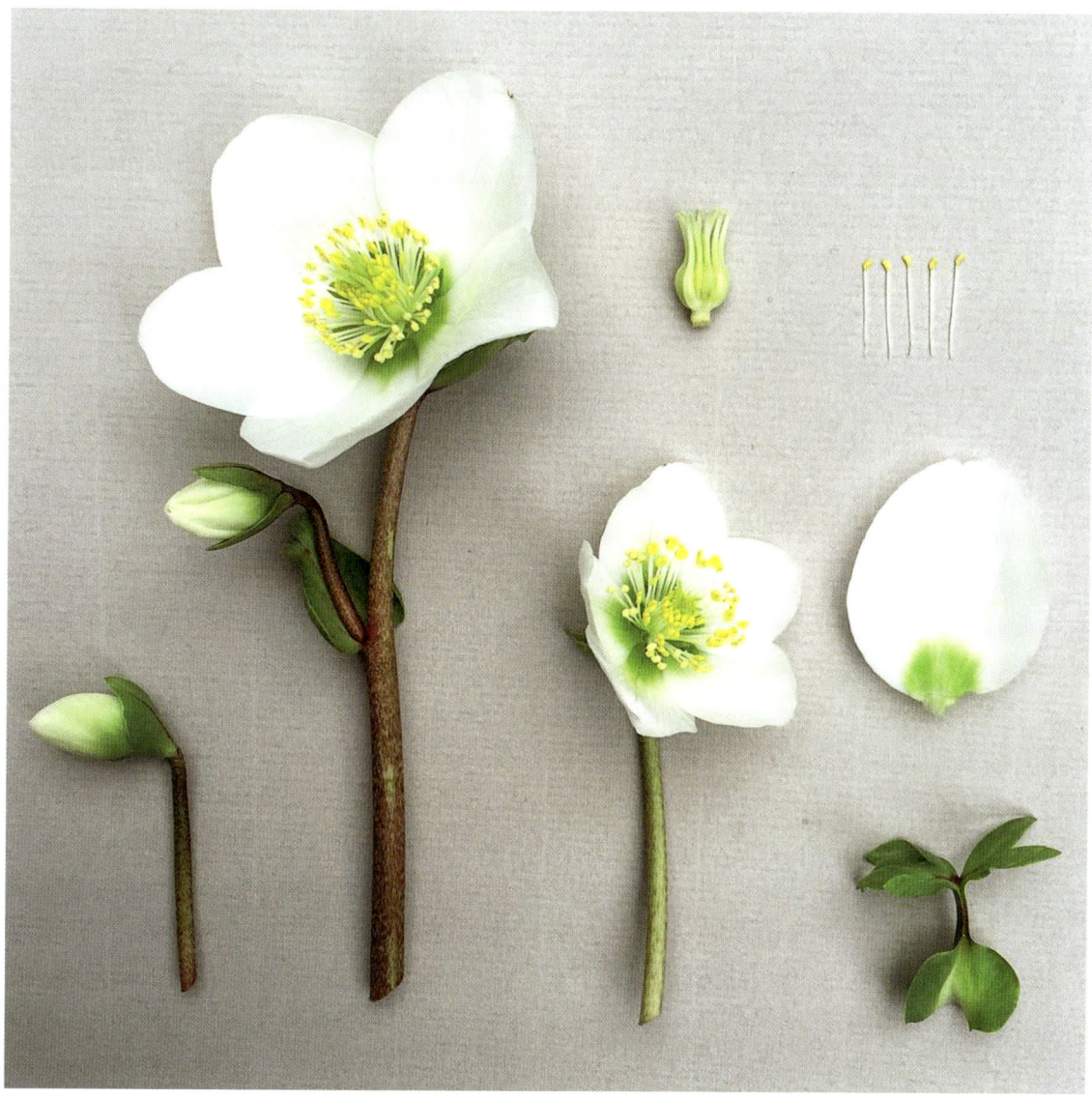

Latin name *Helleborus niger*
Photograph shows Bud, flower with bud, stamens, nectaries, petal, leaves, flower.
Collected London, UK, 2016
Other common name Bear's foot
Family *Ranunculaceae*
Origin Europe
Description A semi-evergreen, perennial that can be up to 30cm in height. Grows from a rhizome.
Leaves Produces dark-green, leathery, pedate leaves.
Flowers Bears cup- or bowl-shaped, pristine white flowers up to 8cm in diameter in mid winter or early spring.

Each one has five sepals and a ring of conspicuous stamens in the centre.
Fruit The fruits are follicles.
Nomenclature The name *Helleborus* is thought to come from the Greek *ellos/hellos*, meaning 'fawn' and *bora* meaning 'food', so would mean 'food for a fawn'. Alternatively, *hele* means to take away, so it could be 'take way food', which refers to the emetic effect of the plant. The epithet *niger*, meaning 'black', describes the plant's roots.
Uses Decorative ornamental. Hellebores are a strong purgative and are used in some homoeopathic remedies.

For the ten years I was based at London's New Covent Garden Market I lived nearby in a little London square with a magical garden at its centre. Laid out by garden designer Dan Pearson in the 1980s, it is now managed by the residents. In mid-winter, Christmas roses — one of the most pristine flowers I know — bloom under the walnut tree.

NIGHT-SCENTED TOBACCO

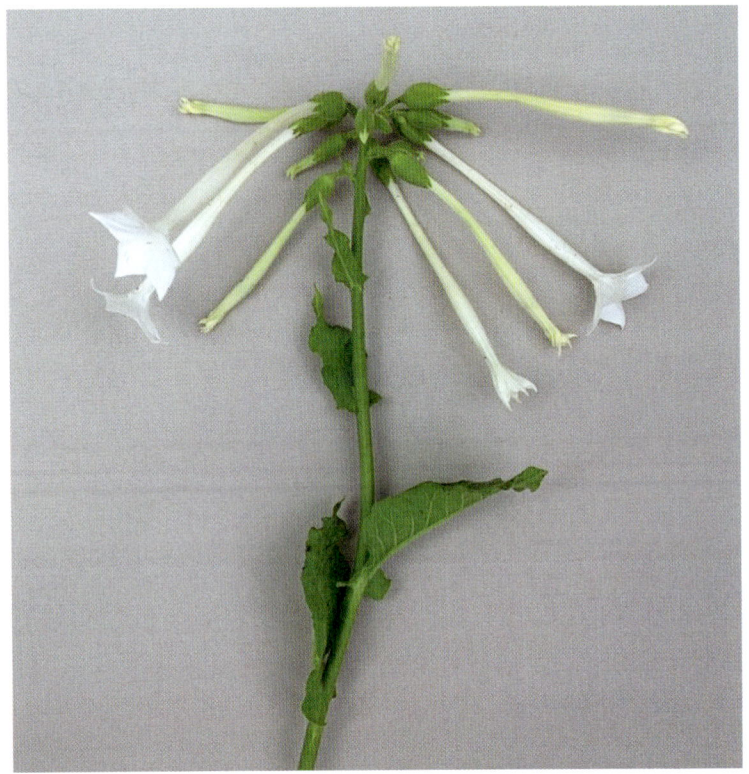

Latin name *Nicotiana sylvestris*
Photographs show Left: raceme of flowers. Bottom left: flower, three buds in various stages of opening, flower, leaf.
Collected London, UK, 2015
Other common name Flowering tobacco
Family *Solanaceae*
Origin Andes regions of Argentina and Bolivia.
Description A branching biennial or short-lived perennial, sometimes grown as a half-hardy annual in colder climates, which can grows to 1.5m high. Understood to be one of the parents of *Nicotiana tabacum*, the plant used in modern tobacco production.
Leaves Forms basal rosette of elliptic, dark-green leaves that can be 30cm long.
Flowers Clusters of fragrant, tubular, white flowers, 7cm long with a face 2cm in diameter, are borne in racemes. Scent is strongest in the evening.
Fruit Not ornamentally significant; the plant forms 12–16mm-long capsules that contain minute brownish seeds.
Nomenclature *Nicotiana* and the word nicotine are named after a French ambassador to Portugal, Jean Nicot (1530–1600), who sent tobacco back to the French court of Catherine de' Medici (1519–89) for medicinal purposes. *Sylvestris*, means 'of or pertaining to a forest or wood'.
Uses Decorative ornamental grown for its night-scented flowers.

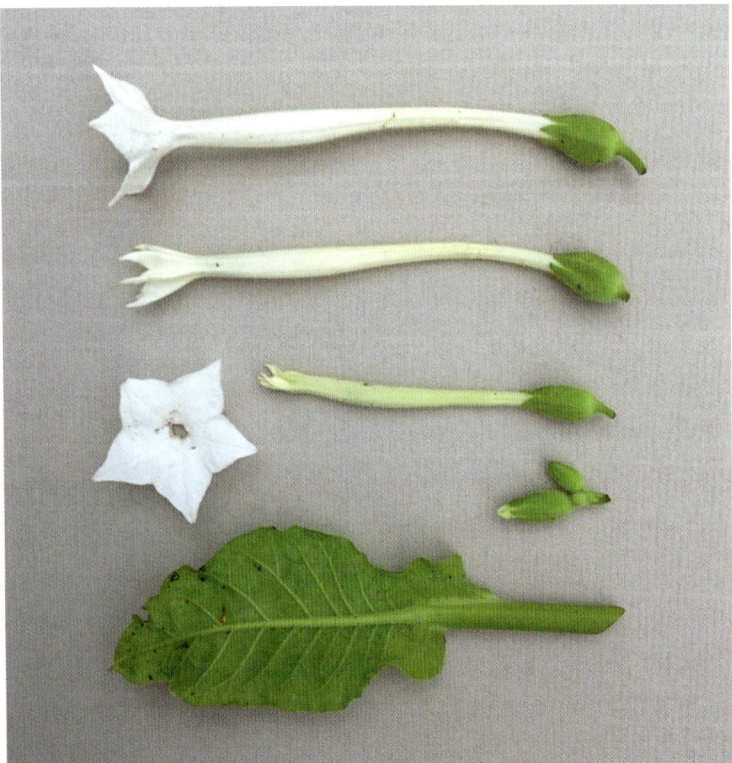

If I crush a leaf of this plant and hold it to my nose I am taken back to the tobacco fields of Birkdale, my cousin's farm in Umvukwes (now Mvurwi) – to my mind it is one of the prettiest places in Africa. We spent many happy weekends scrambling up kopjes (rocky outcrops), searching for game, and camping by the river. We had great adventures there and loved Birkdale. Sadly, the farm was compulsorily acquired under the Zimbabwean land acquisition of 2002 with no compensation.

TEA PLANT

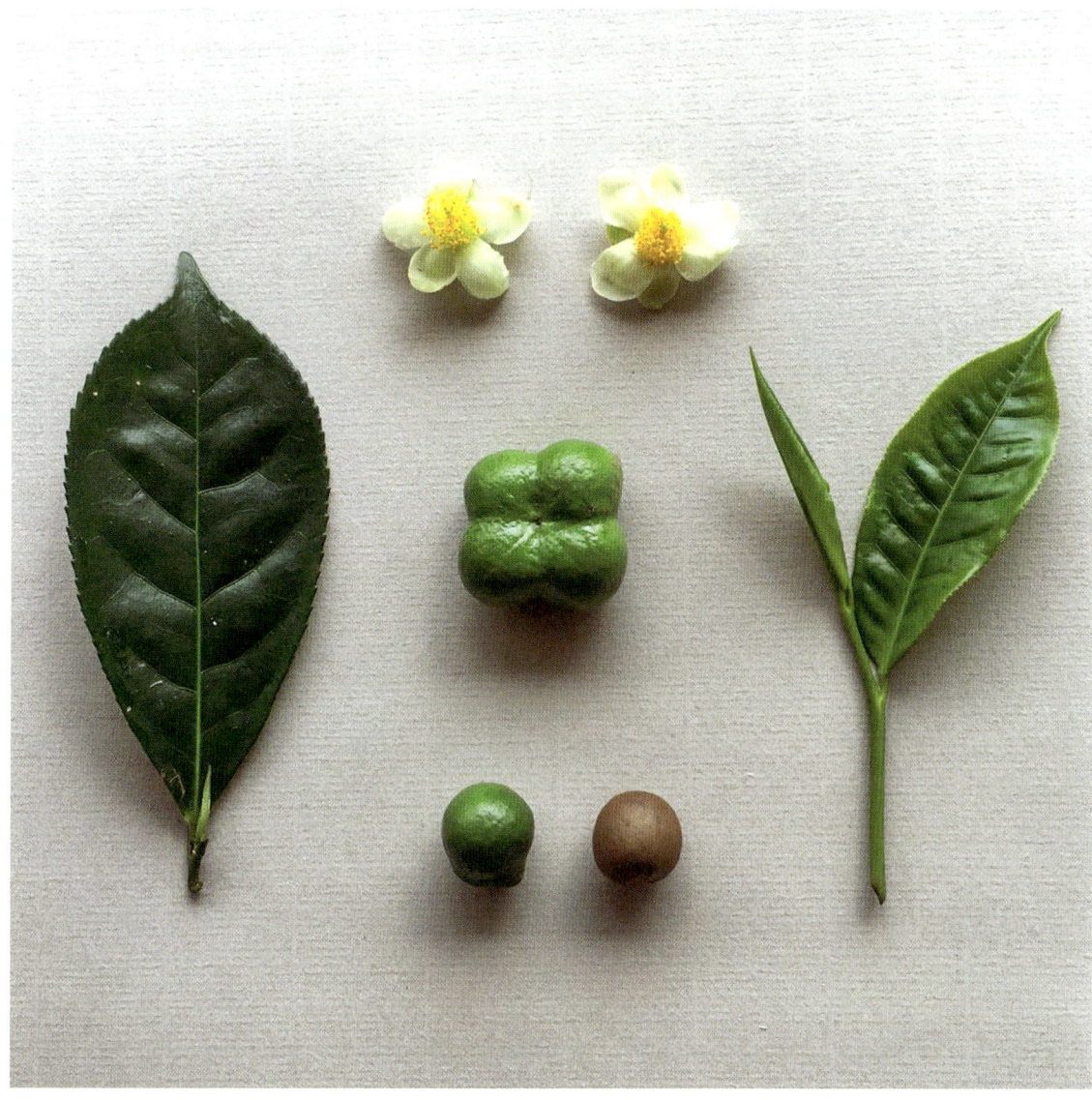

Latin name *Camellia sinensis*
Photograph shows Mature leaf, flowers, young leaves, seeds (young and mature), seed capsule.
Collected Aberfolye Lodge, Honde Valley, Zimbabwe, 2016
Other common names Camellia, Tea shrub
Family *Theaceae*
Origin India and Southeast Asia.
Description A medium-sized, evergreen, woody, autumn/winter-flowering shrub, that is usually trimmed to a maximum height of 2m, but can grow to 4m.

Leaves Dark-green, oval leaves, with slightly dentate margins, are pointed at the tip, and grow to 10cm long.
Flowers White, fragrant, five-petalled flowers are up to 4cm in diameter.
Fruit Forms capsules with one to four lobes each of which contains one seed.
Nomenclature Carl Linnaeus named the genus *Camellia* after Moravian botanist Joseph Kamel (1661–1706). *Sinensis*, means 'Chinese' or 'China'.
Uses Often grown as a cash crop at high altitudes, the dried leaves are used commercially to make tea.

I collected this specimen at Aberfoyle Lodge, a tea plantation in the stunning Honde valley in the Zimbabwean Eastern Highlands.

WHITE PASSION FLOWER

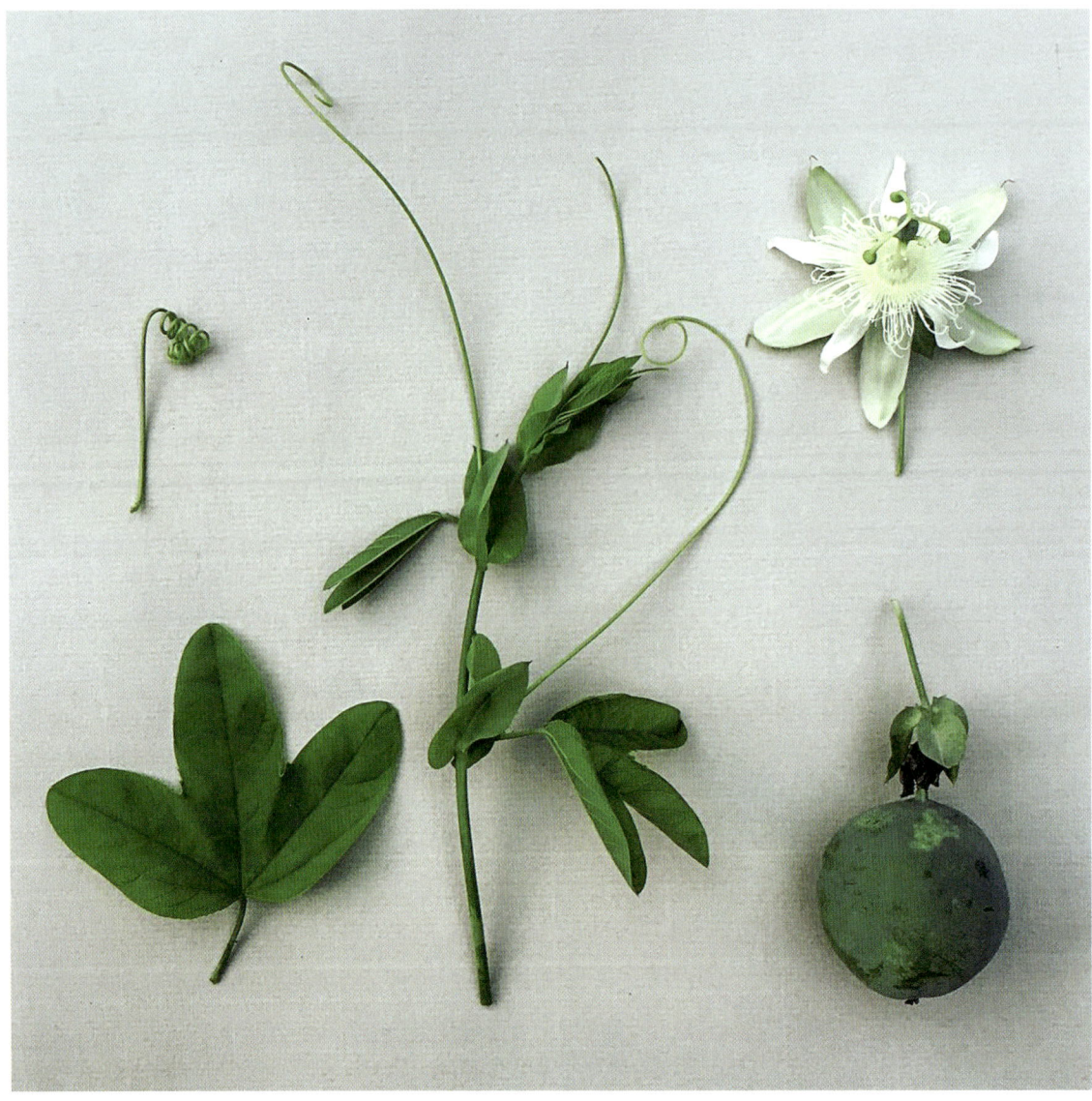

Latin name *Passiflora subpeltata*
Photograph shows Leaf, tendril, section of vine, flower, fruit.
Collected La Rochelle, Penhalonga, Zimbabwe, 2015
Family *Passifloraceae*
Origin Mexico, Central America, and tropical South America.
Description Herbaceous, summer-flowering, vine climber, 6–9m in height.
Leaves The three-lobed leaves are pale green with bluish-green undersides.
Flowers Bears single, white flowers about 4–5.5cm in diameter in the leaf forks.
Fruit Egg-shaped fruits are 4cm long and have leather-like skin. Pale green at first, they turn yellowish when mature.

Nomenclature *Passiflora* means 'passion flower'. Spanish missionaries in South America named them 'Passion Flowers' as the parts of the plant symbolised the crucifixion of Jesus. The five petals and five sepals represent 10 apostles (excluding Judas and Peter), the three stigma are the nails that held Jesus on to the cross and the five anthers his wounds. The tendrils are said to resemble the whips used to flagellate Christ, and the flower's filaments depict the crown of thorns. *Subpeltata* is from the Latin *pelate* meaning shield-shaped, and *sub* refers to the petiole attached to the lower surface of the leaf.
Uses None found.

This species arrived in Zimbabwe in a seed packet along with other plants including *Lillium formosanum*, *Verbena bonariensis*, and *Cosmos bipinnatus*. This passion flower has since naturalised and grows wild across the Zimbabwean countryside.

GRANADILLA

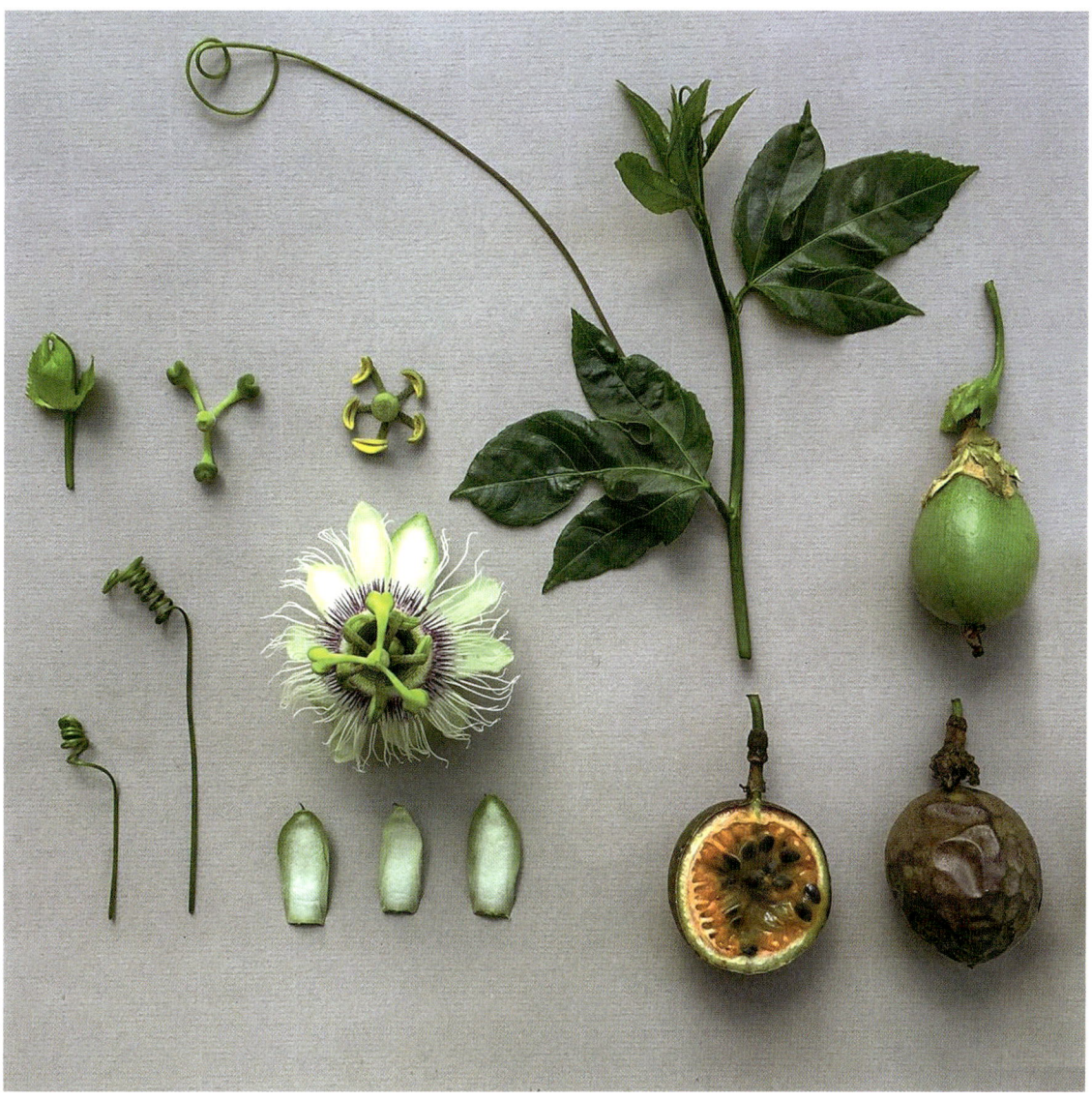

Latin name *Passiflora edulis*
Photograph shows Leaflet, stigma, anthers, vine with tendrils, fruit, mature fruit, section of fruit, petals, flower, tendrils.
Collected Harare, Zimbabwe, 2016
Other common name Passion fruit. In the Antipodes, the fruit of this plant are called 'granadilla' and the yellow fruit of *P. ligularis* are known as passion fruit.
Family *Passifloraceae*
Origin South America – southern Brazil, Paraguay to northern Argentina.
Description A vigorous, evergreen, summer-flowering, climbing vine that clings by tendrils to almost anything. It can grow 4.7–6m a year once

established, but needs strong support.
Leaves The three-lobed leaves with serrate margins are 7–20cm long and glossy above, but dull below. There are two small glands on the stalks. Young stems are tinged with red or purple.
Flowers Solitary flowers, up to 7cm in diameter, have white petals, and a corona with four to five rows of white filaments, with purple bases, up to 2.5cm long.
Fruit Ovoid to spherical fruit are 4–5cm in diameter. They become greenish-yellow, then purplish as they mature.
Nomenclature *Passiflora* means 'Passion flower' (see opposite); *edulis* is 'edible'.
Uses Grown as a fruit crop. It is used in traditional South American remedies.

My grandfather loved this fruit and planted row upon row of granadilla vines in the vegetable garden at Maduma, training them up chicken wire. My mother, whose creative eye I have often turned to while struggling to resolve a botanical layout, suggested that I photograph the flowers and fruit of this vine.

ANGEL WING JASMINE

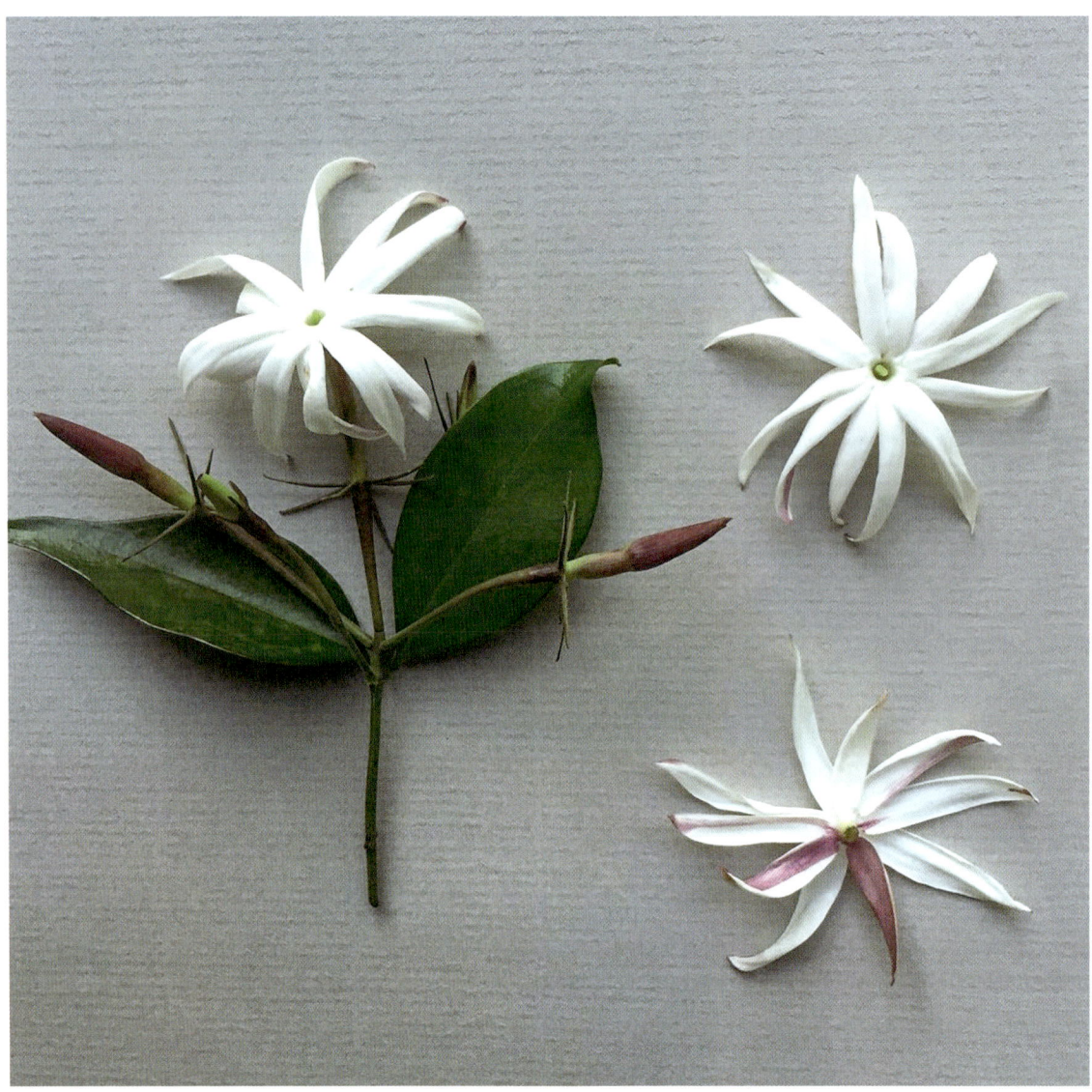

Latin name *Jasminum nitidum*
Photograph shows Stem (with leaves, flower, and buds), flower from above, underside of flower.
Collected Harare, Zimbabwe, 2016
Other common names Star jasmine, Shining jasmine
Family *Oleaceae*
Origin Admiralty Islands, Papua New Guinea.
Description An evergreen or semi-evergreen, fast-growing shrub or vine that grows up to 6m tall.
Leaves Produces opposite, elliptical or lance-shaped, glossy, leaves that grow up to 5cm long.
Flowers From late spring to autumn, plant bears pinkish buds that open into white, pinwheel-shaped, fragrant flowers that can be up to 5cm across. Underside of petals can be pink.
Fruit Black globose berries.
Nomenclature *Jasminum* comes from old French *jasmin*, from Arabic *yāsamīn*, from Persian *yāsmīn*; *nitidum* is from Latin meaning 'shiny', and refers to the glossy leaves.
Uses Decorative ornamental.

Bearing sprays of star-shaped, scented flowers, the perfume of this jasmine lingers deliciously at dusk.

WINTER JASMINE

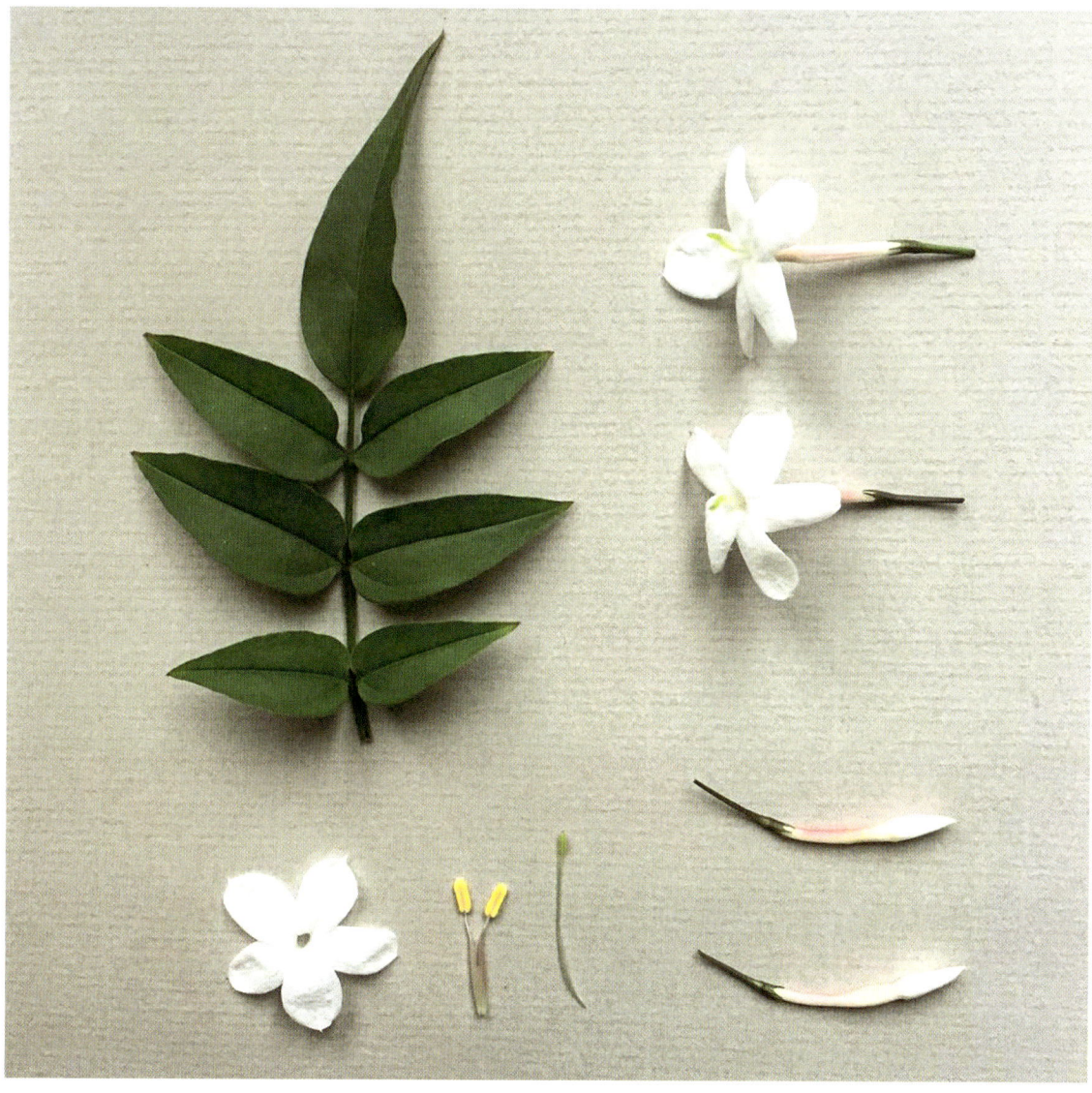

Latin name *Jasminum polyanthum*
Photograph shows Leaf, side view of flowers, buds, pistil, stamens, open flower from above.
Collected London, UK, 2015
Other common name Pink jasmine
Family *Oleaceae*
Origin South-western China
Description Vigorous, evergreen, twining climber that bears masses of intensely fragrant flowers in late winter or early spring. Grows up to 6m.
Leaves Opposite, pinnately compound leaves (up to 10cm long) have two to three pairs of pointed leaflets and one longer terminal one. They are dark green on top with a paler underside.
Flowers Scented, white, long-tubed, star-shaped flowers (2cm in diameter) are borne in dense clusters.
Fruit Plant occasionally produces small, black, globular fruit.
Nomenclature *Jasminum* comes from old French *jasmin*, from Arabic *yāsamīn*, from Persian *yāsmīn*. *Polyanthum* originates in the Greek words *polús* meaning 'many' and *anthos*, or 'flower'.
Uses Decorative ornamental.

In Zimbabwe, the jasmine starts flowering in July, one of the coldest and driest months of the year. Whenever I smell the flowers of this species, anywhere in the world, I am transported back to a dusty African winter.

DELICIOUS MONSTER

Latin name *Monstera deliciosa*
Photograph shows Flower, unripe fruit.
Collected Maduma, Harare, Zimbabwe, 2014
Common names Fruit salad plant, Swiss cheese plant, Split-leaf philodendron.
Family *Araceae*
Origin Tropical rainforests of southern Mexico to Central America.
Description An epiphytic vine with thick aerial roots that it uses for climbing up tree trunks; it can grow to more than 9m.
Leaves The dark-green, leathery leaves can be up to 90cm across. The edges of the juvenile leaves are unbroken, but as the plant matures the leaf edges become deeply cut and elliptic holes develop along the midribs. Flattened petioles grow up to 1m in length.
Flowers Bears a whitish-green inflorescence, or spadix, in a white spathe up to 30cm long.
Fruit A compound fruit up to 40cm long and 1cm wide is born on the spathe. Fruit drops off when it eventually ripens.
Nomenclature *Monstera* may come from the Latin *monstrum*, meaning 'monster' referring to the sheer size of this plant. The epithet, *deliciosa*, means delicious, which describes the edible fruit that only forms in the plant's natural habitat.
Uses Grown as a decorative ornamental, but the fruit that develops in the wild is also edible.

As its Latin name suggests, the fruit of this tropical plant is utterly delicious and reminiscent of custard apples and pineapple. When the fruit is ripe (which can take up to a year) its hexagonal scales start to fall away, revealing delicious flesh that exudes an intense, tropical perfume.

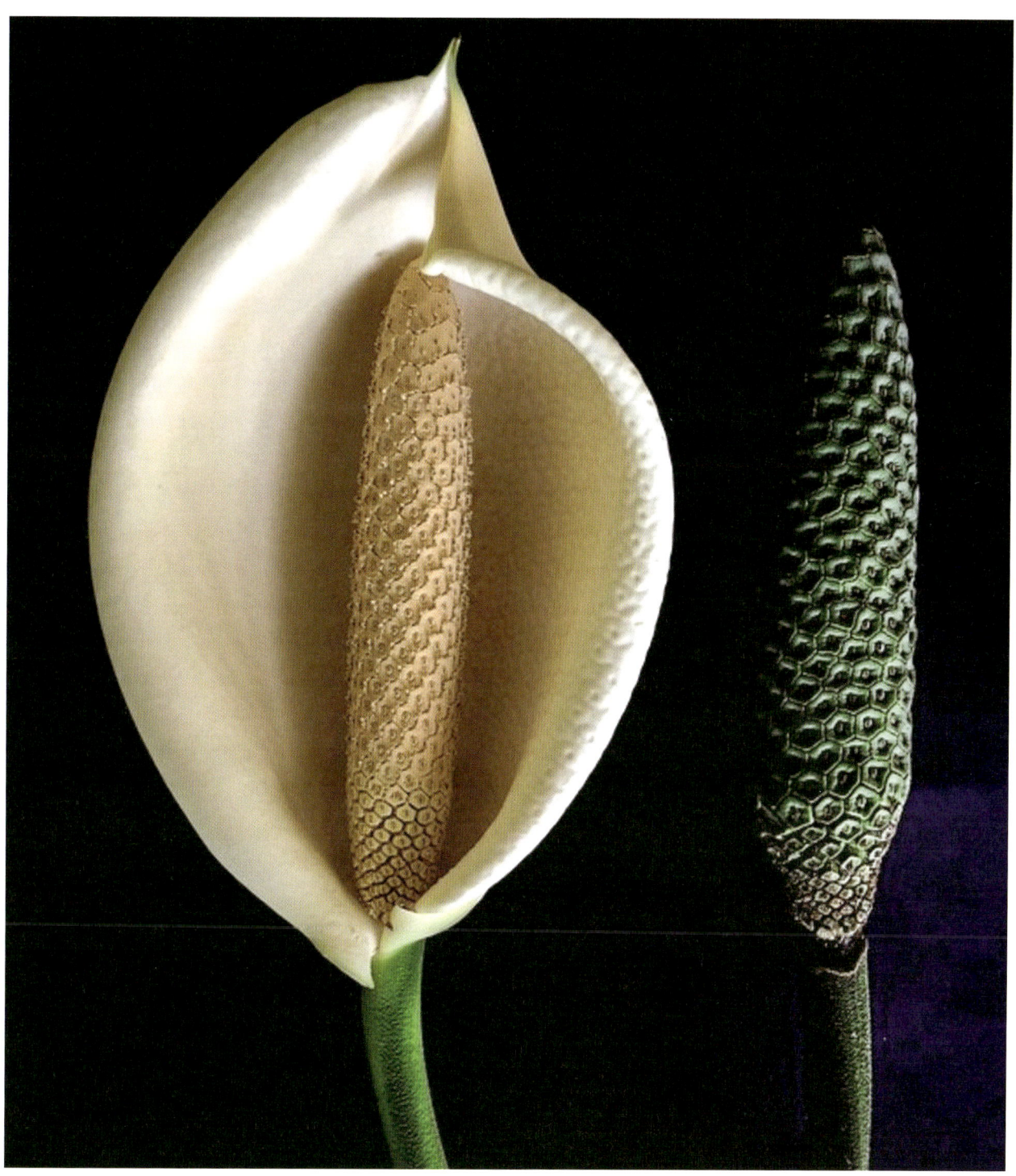

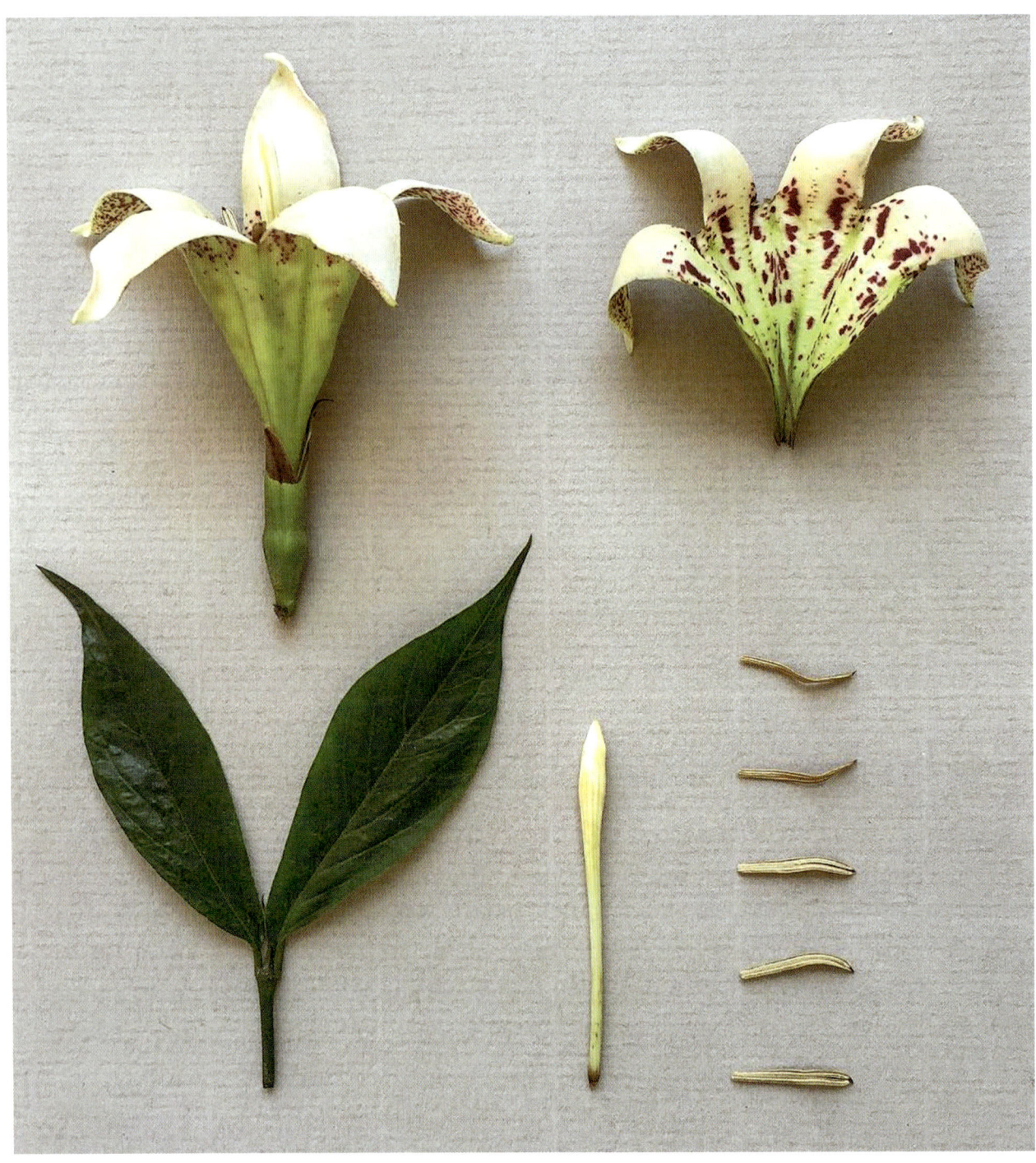

CAPE GARDENIA

Latin name *Rothmannia capensis*
Photographs show Right: single flower. Opposite: side view of flower, half flower, anthers, stamen, leaves.
Collected Maduma, Harare, Zimbabwe, 2014
Other common names Wild gardenia, Common rothmannia
Family *Rubiaceae*
Origin Forests and rocky areas of South Africa at elevations of up to 1,900m.
Description Slow-growing, summer-flowering tree that reaches a height of 10m in woodland and 20m in forests.
Leaves Glossy, green leaves are 7–10cm long with domatia (arthropod shelters) in the axils.
Flowers The fragrant, bell-like flowers are about 80mm long with reddish streaks down the flower tube.
Fruit Produces round, 70mm fruit that is green at first, becoming brown with grooves on the skin as it ripens. The pulp contains the flat seeds.
Nomenclature *Rothmannia* was named by botanist Carl Thunberg (1743–1828) after his friend and co-pupil of Linnaeus, the Swedish botanist Dr Georgius Rothman (1739–78). The Latin word *capensis* means 'from the Cape'.
Uses Wood is pliable and carved to form instrument handles and durable household utensils including wooden spoons. It makes excellent for firewood. The fruit is both edible and palatable.

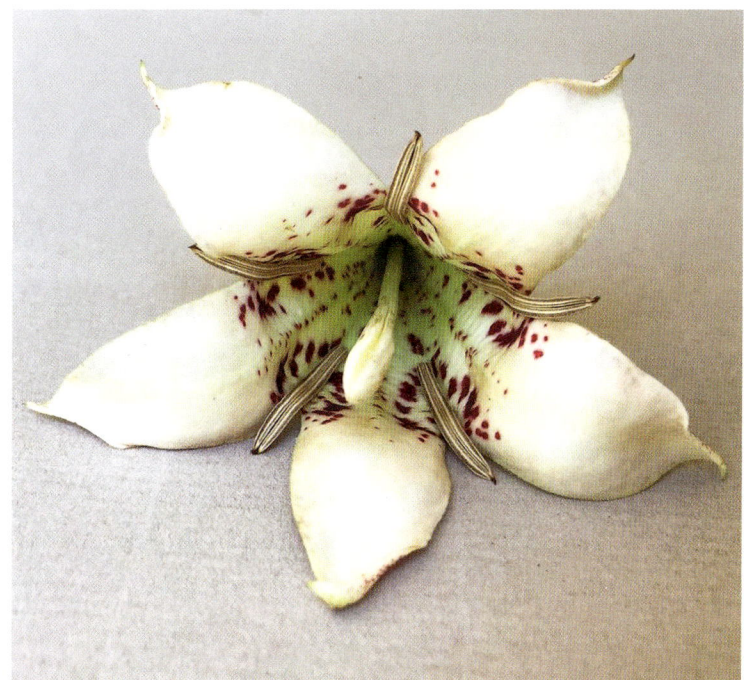

After decades of planting exotics, there is a strong movement in Zimbabwe to plant native species as they are more ecologically sound, generally require less water, and are more suited to the climate. I collected the flowers from this indigenous tree that I planted to attract birds to Maduma's garden.

GARDENIA

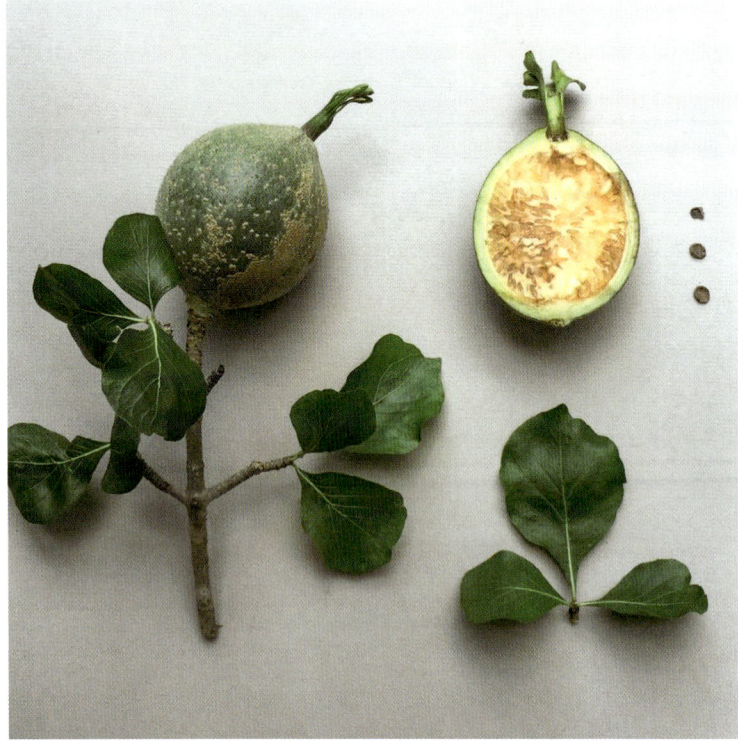

Latin name Left: *Gardenia thunbergia*
Below: *Gardenia posquerioides*
Photographs show Left: stem with leaves and fruit, section of unripe fruit, seeds, leaf. Below: flower.
Collected Left: National Botanical Gardens, Harare, Zimbabwe, 2016.
Below: La Rochelle, Penhalonga, Zimbabwe, 2015.
Other common names Forest gardenia, Tree gardenia
Family *Rubiaceae*
Origin South Africa, eastern Kenya, Tanzania, and Zimbabwe.
Description An evergreen shrub or small tree that can be 2–5m high.
Leaves Glossy, light-green leaves are borne in whorls of three to four per stem.
Flowers Large, showy, white blooms, 7cm in diameter, emerge from green buds and do not yellow as they age. Each one is borne on a conspicuous corolla, and has eight petals. At their most fragrant at night, flowers are pollinated by hawkmoths.
Fruit Hard, woody, egg-shaped fruits grow up to 8cm long. Greyish in colour at first, they develop some encrustation as they mature and can remain on the plant for many months.
Nomenclature The genus *Gardenia* is named after a Scottish-born American naturalist Dr Alexander Garden (1730–91), and *thunbergia* is named after Swedish botanist Carl Thunberg (1743–1828), who travelled in South Africa.
Uses In southern Africa, the roots, root bark, leaves, and latex from this tree are used medicinally. The dense, hardwood is used to make small tools, buttons, and handles.

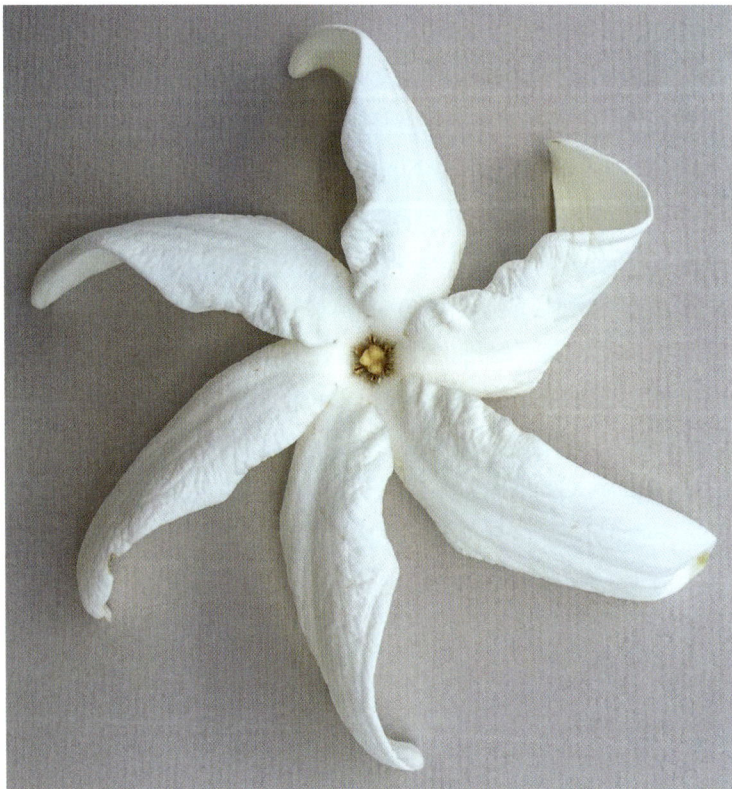

The Gardenia is an interesting example of mutualism between species. The robust fruit do not split open naturally. For the seeds to be released the fruit must first pass through the stomachs of the animals that eat them, so it relies on elephants, buffalo, and large antelope to reproduce.

GREAT WHITE CHERRY

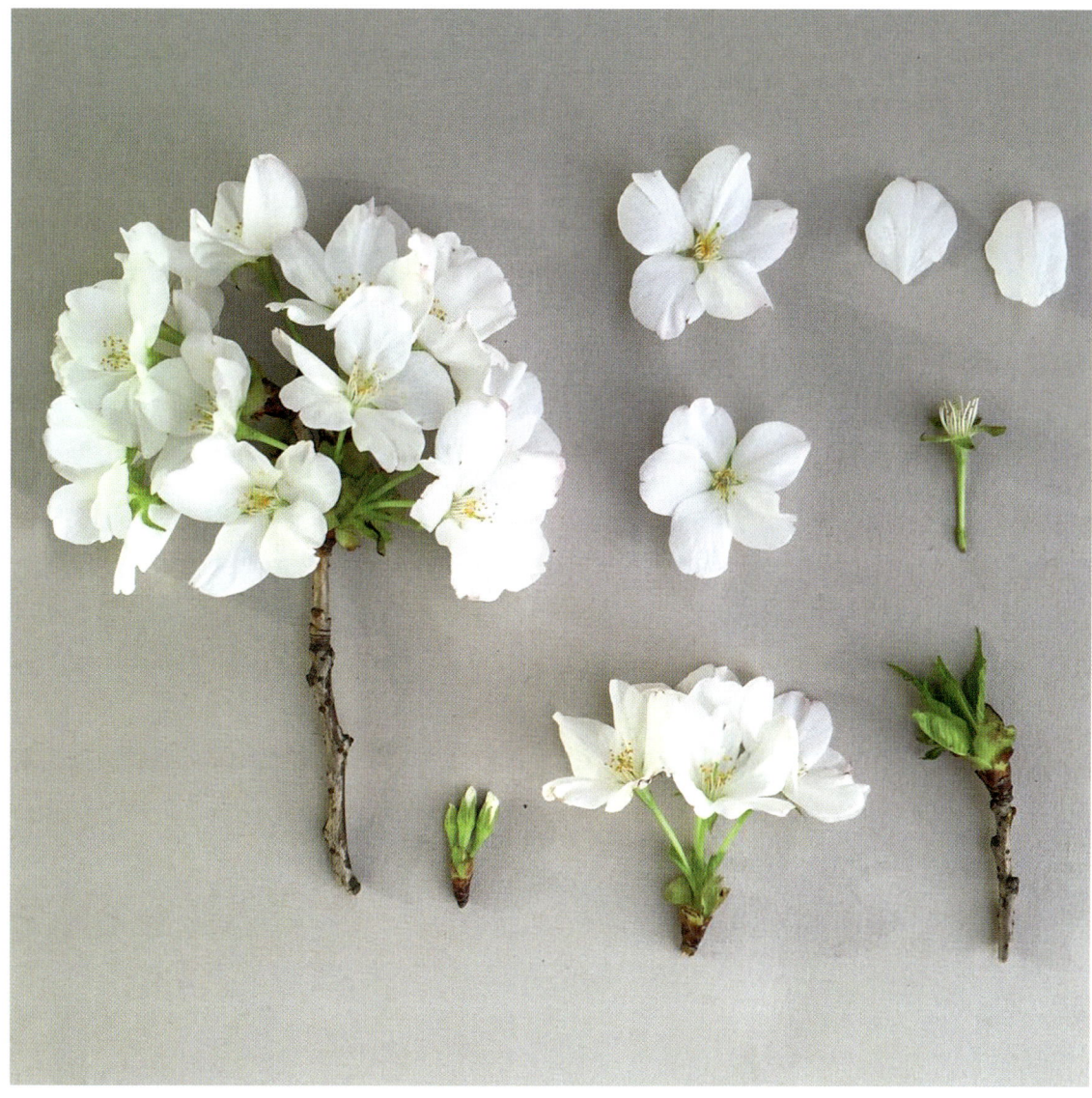

Latin name *Prunus* 'Tai-haku'
Photograph shows Flower cluster, single flowers, petals, peduncle showing anthers, emerging leaf shoot, small flower cluster, buds.
Collected Beijing, China, 2016
Other common name Flowering cherry
Family *Rosaceae*
Origin Japan
Description A medium-sized, deciduous tree growing up to 6m with a broad and spreading crown. Both leaves and flowers appear at the same time.

Leaves Young leaves are deep bronze, then green and turn yellow and orange in autumn. Mature leaves have serrated margins and can be up to 15cm long.
Flowers Large, pure-white single flowers up to 6cm in diameter are borne in clusters in spring with the young leaves.
Fruit None.
Nomenclature *Prunus* is from Latin, and literally means 'plum tree'. The cultivar name 'Tai-haku' is Japanese for 'big white flower'.
Uses Graceful ornamental tree.

An ancient Japanese tree, *Prunus* 'Tai-haku' became extinct in its native country. This beautiful white cherry species was revived from a single specimen found in a garden in Sussex, England.

LOEBNER MAGNOLIA

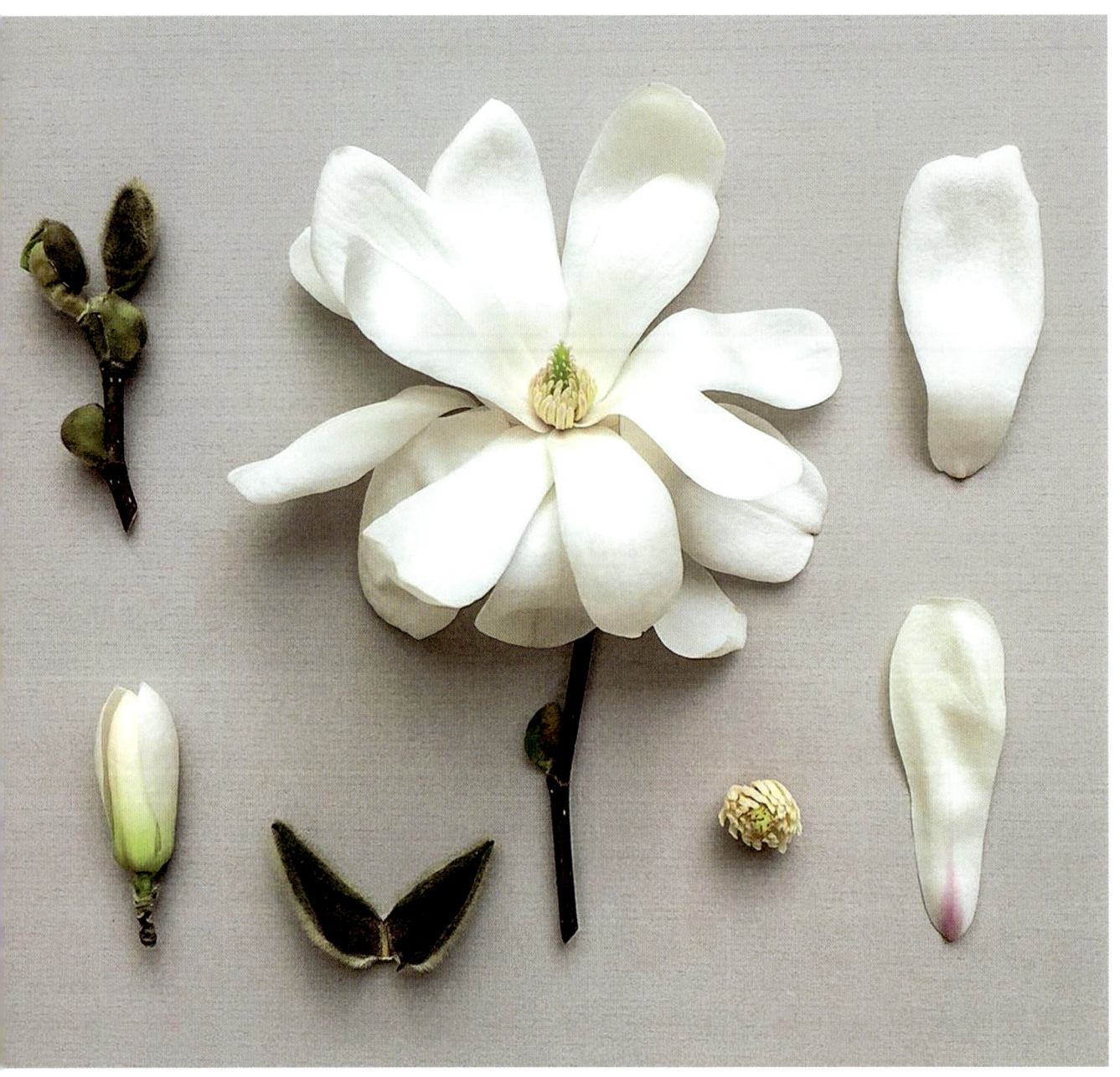

At the top of London's Kings Road, Chelsea, grow a couple of Loebner magnolias, which are an utter delight when they are smothered in flowers in early spring.

STAR MAGNOLIA

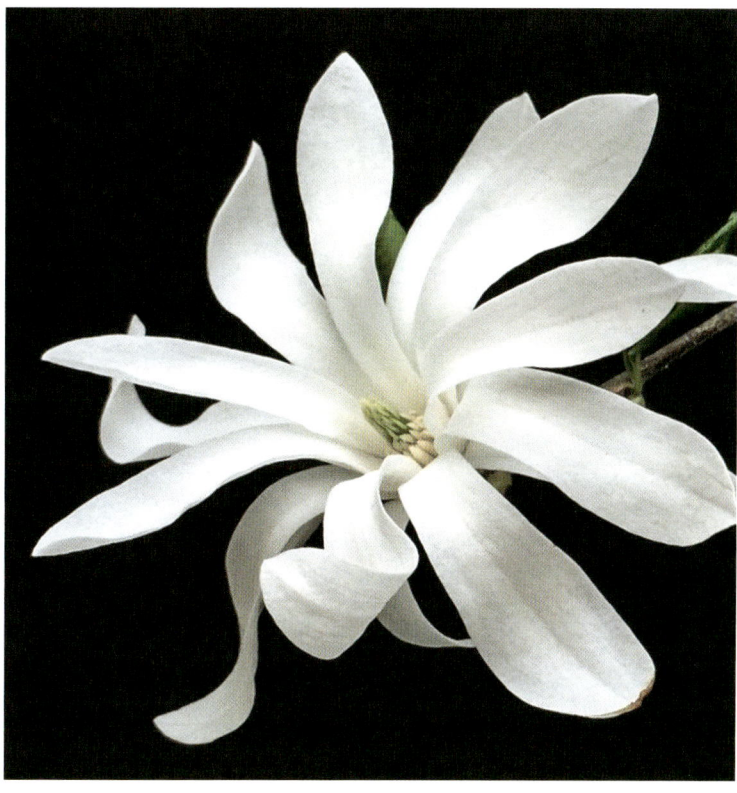

Latin name *Magnolia* x *loebneri* 'Merrill'
Photograph (opposite) shows Flower
bud, leaf buds, flower, petals, carpel and
stamens, bud casing.
Collected London, UK, 2015
Family *Magnoliaceae*
Origin Hybrid bred in Germany. Parent
plants *M. stellata* (right) and
M. kobbus both originate in Japan.
Description A small, deciduous, hybrid
grown as a shrub or tree with a rounded
crown, which can reach 9m tall.
Leaves Obovate, green leaves are up
to 12cm long.
Flowers Bears fragrant, star-like,
white flowers (10–15cm across) with
10 to 15 petals, that open before the
leaves appear.
Fruit This hybrid does not produce
any fruit.
Nomenclature The genus is named
after the French botanist Pierre Magnol
(1683–1715). This hybrid, which first
flowered in 1917, is named after the
German plant breeder Max Löbner who
created this hybrid magnolia.
Uses Decorative ornamental.

When staying with a friend in her cottage in Oxfordshire, I saw a
beautiful *Magnolia stellata* blooming in an adjacent garden. Ever the
botanical poacher – and knowing the neighbours were away –
I hopped the low wall and collected this flower to photograph.

Latin name *Magnolia stellata*
Photograph shows Open flower
Collected Oxfordshire, UK, 2016
Family *Magnoliaceae*
Origin Japan
Description A slow-growing, deciduous,
spring-flowering shrub or small tree of
broadly rounded habit, which grows up
to 2.5m tall, with a spread of 2.5–4m.
Leaves Oblong leaves that measure
10cm long are bright green in summer
and turn yellow to bronze in autumn.
Flowers The large, white flowers
measuring 10cm across, have up to 18
narrowly oblong, spreading white tepals
in early spring before leaves appear.
Fruit Reddish green, knobby aggregate
fruit measure 5cm long. Mature fruit
opens to reveal orange-red seeds.
Nomenclature *Stellata*, is from the Latin
meaning 'star', and describes the flowers.
Uses Decorative ornamental.

TULIP MAGNOLIA 'LENNEI ALBA'

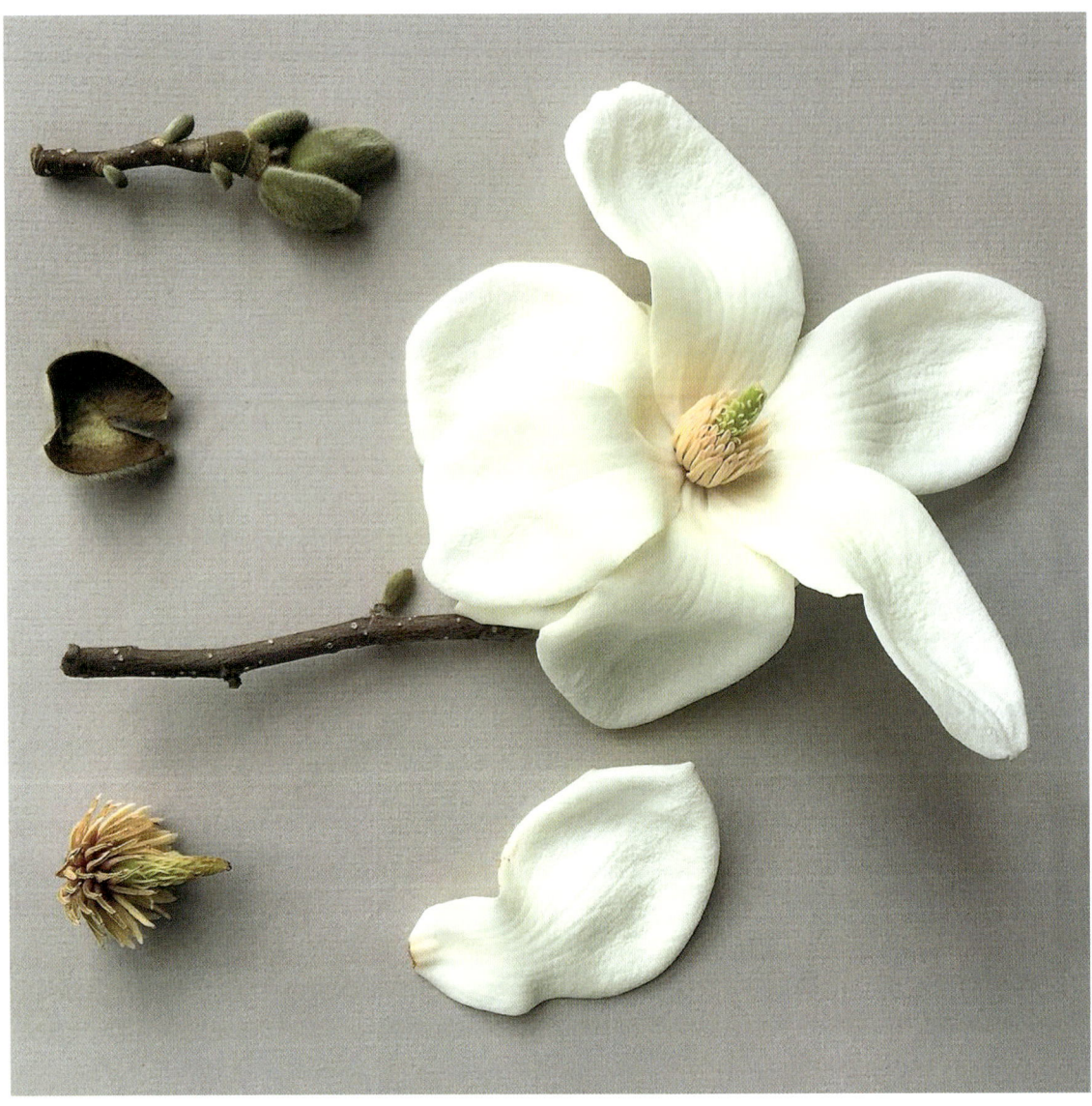

Latin name *Magnolia* x *soulangeana* 'Lennei Alba'
Photograph shows Fruit casing, buds on a stem, flower, petal, corona.
Collected Beijing, China, 2015
Other common names Chinese magnolia, Saucer magnolia
Family *Magnoliaceae*
Origin Parents of the hybrid (*M. denudata* and *M. liliflora*) are native to China.
Description A multi-stemmed, deciduous, large shrub or small tree with a height of 8m and a spread of 6–9m.

Leaves Obovate, alternate, dark-green, hairy leaves up to 20cm long appear after flowers.
Flowers Bears goblet-shaped flowers 10–20cm across in spring.
Fruit Elongated fruit are 7cm long, and house small red seeds.
Nomenclature *M.* x *soulangeana* hybrid was bred by E Soulange-Bodin (page 104). This cultivar was bred by Swiss botanist Karl Otto Froebel (1844–1906) in 1905.
Use Decorative ornamental.

I spotted this tulip magnolia cultivar tree flowering in the 798 Art District in Beijing, China, on a trip there in 2015, and made a nocturnal foray to collect this specimen. The large, creamy-white flowers have thick petals and a lemony fragrance.

SOUTHERN MAGNOLIA

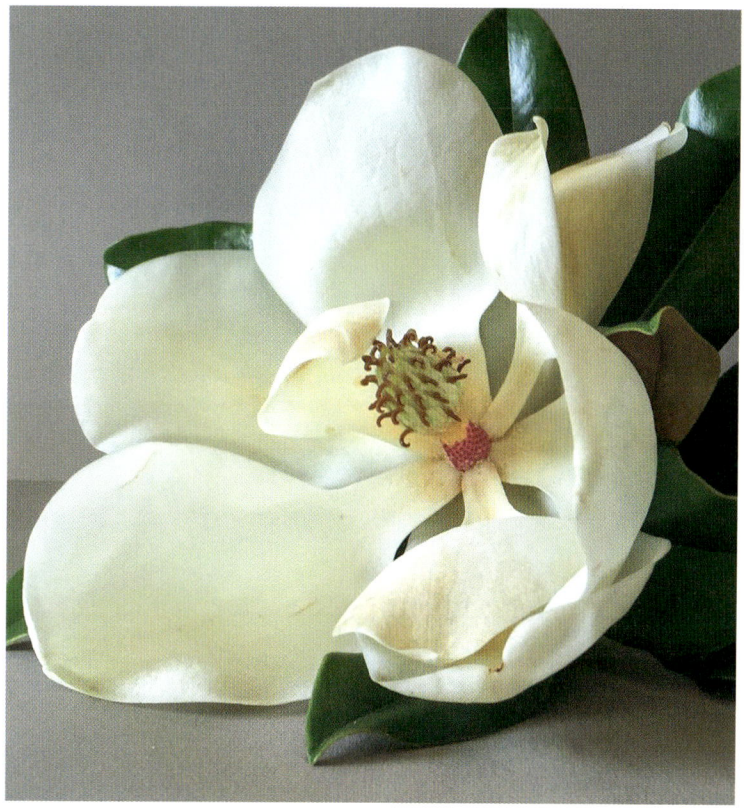

Latin name *Magnolia grandiflora*
Photograph shows Flower head
Collected Vicenza, Italy, 2013
Other common name Bull Bay magnolia
Family *Magnoliaceae*
Origin Lowland subtropical forests of southeastern USA.
Description Large spring-flowering, evergreen tree that grows up to 30m tall.
Leaves Simple, broadly ovate, leathery, stiff leaves are dark-green with bronze undersides and are 12–20cm long and 6–12cm wide.

Flowers Bears large, showy, white, lightly lemon-scented flowers, 30cm in diameter, that have six to 12 waxy petals.
Fruit Forms cone-shaped, woody, aggregate fruits, 5–10cm long. When fruit is mature, shiny, red, kidney-like seeds hang from filamentous threads.
Nomenclature The genus *Magnolia* is named after the French botanist, Pierre Magnol (1638–1715). The species name *grandiflora*, means 'large flowered' and refers to the spectacular blooms.
Uses Decorative ornamental tree.

Oscar Wilde describes this tree beautifully in *The Birthday of the Infanta*, saying, 'The magnolia trees opened their great globe-like blossoms of folded ivory, and filled the air with a sweet heavy perfume'. I came across this enormous *M. grandifora* tree in flower in the middle of a field.

SPIDER LILY

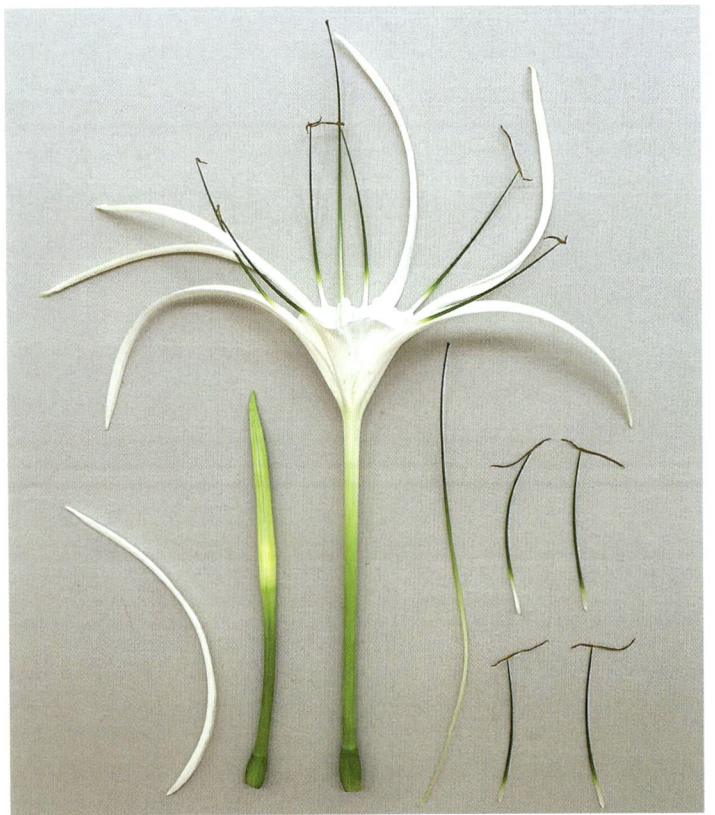

Latin name *Hymenocallis littoralis*
Photographs show Left: tepal, bud, flower, pistil, stamens. Below left: flower head from above.
Collected Left: Langkwai, Malaysia, 2016
Below left: Harare, Zimbabwe, 2015
Family *Amaryllidaceae*
Distribution Warmer, coastal regions of Central and South America.
Origin A vigorous, evergreen, species that grows to 60–70cm in height from a bulb.
Leaves Forms narrow, glossy, dark-green, strap-like, arching leaves that can be 60cm long.
Flowers Bears large, fragrant, white, tubular flowers, 14–17cm long, on thick, flattened stems. Long, slender segments radiate from a round centre, giving the bloom a spider-like appearance – hence the common name.
Fruit Forms loculicidal capsules, with two seeds in each locule.
Nomenclature The genus name *Hymenocallis*, or 'membraned beauty' is from Greek words *hymen*, meaning 'membrane', and *kallos*, meaning 'beautiful', referencing the flower's filament cup. *Littoralis* or 'of the shore', describes the plant's natural habitat.
Uses Decorative ornamental.

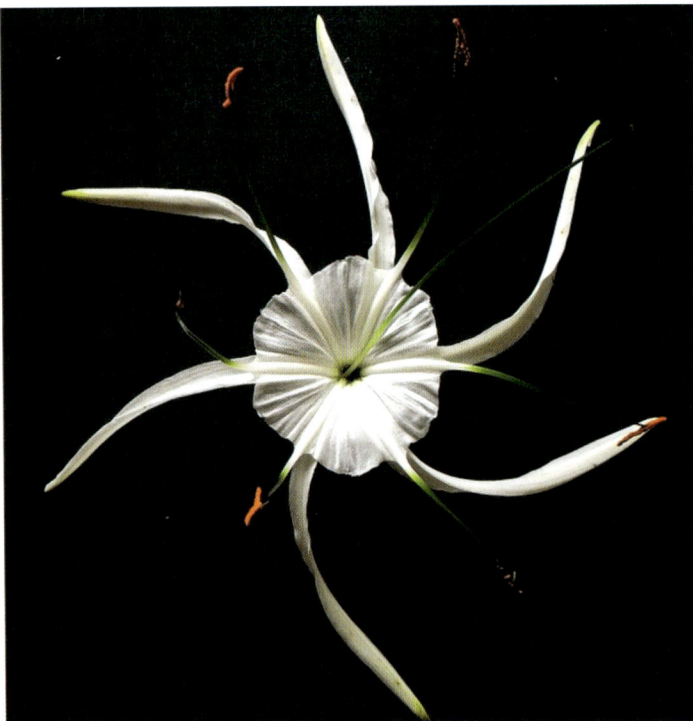

The spidery white flowers of this plant have a pretty scent that is strongest after dusk.

KALAHARI BAUHINIA

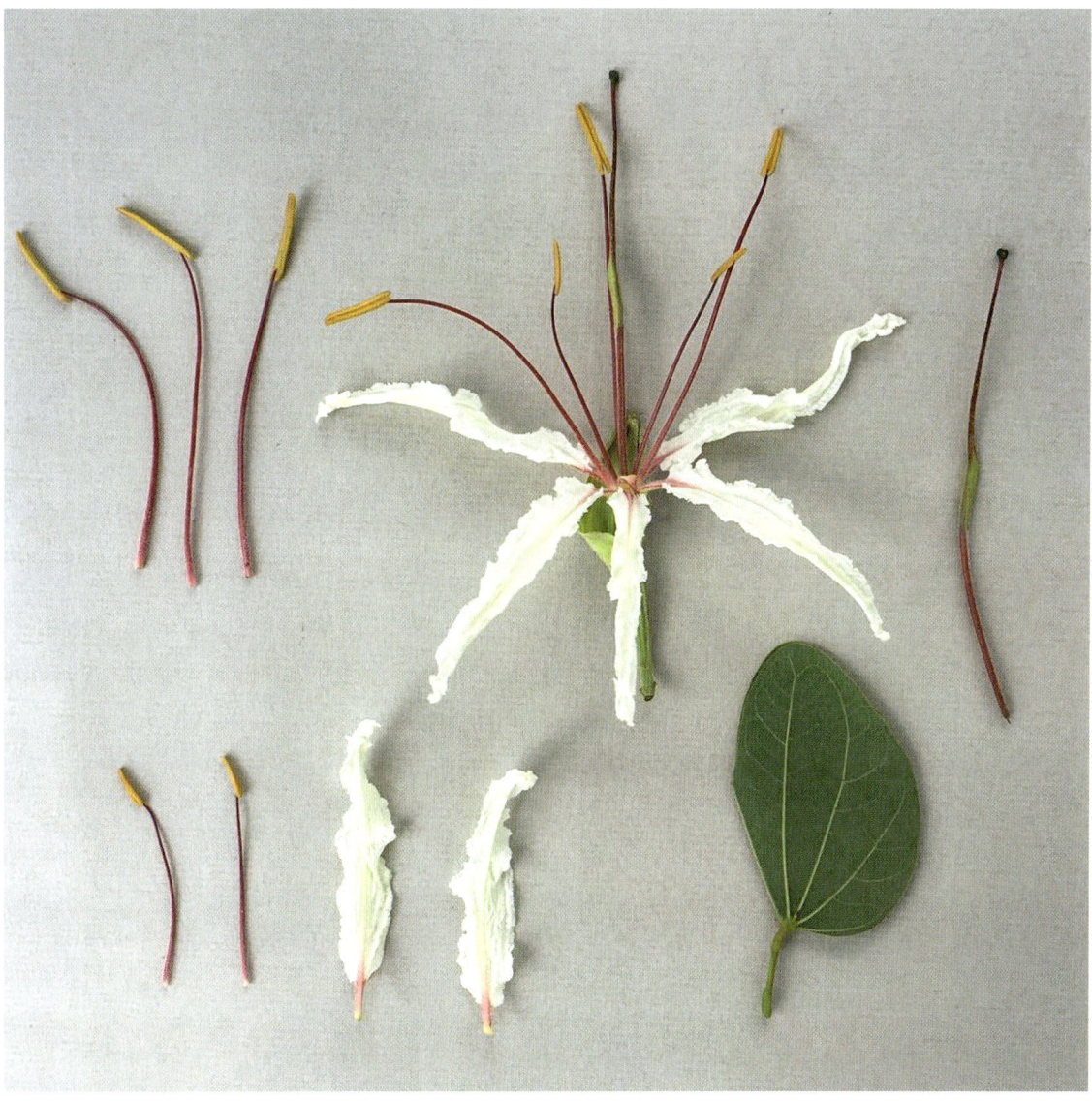

Latin name *Bauhinia petersiana*
Photograph shows Anthers, flower head, stigma, leaf, petals, anthers.
Collected Mutare, Zimbabwe, 2015
Family *Fabaceae*
Origin At altitudes in Tanzania, DRC (Democratic Republic of the Congo), Zambia, Malawi, Mozambique, Zimbabwe, Botswana, South Africa.
Description Large summer-flowering, deciduous bush or small tree that grows up to 2m.
Leaves Produces alternate, simple, 2-lobed leaves; lobes are elliptical to ovate or rounded.
Flowers Large, fragrant, showy terminal clusters have long, narrow, snow-white petals and, usually, wrinkled, reddish stamens and style.
Fruit Woody, dehiscent, flattened pods contain five or six seeds.
Nomenclature The genus *Bauhinia* is named after Swiss–French botanist brothers Johann (1541–1613) and Gaspard (1560–1624) Bauhin. *Petersiana* is named after German zoologist Professor Wilhelm Peters (1815–83), who collected it in Mozambique in the mid-19th century.
Uses Many medicinal uses. Macerated roots of the plant, can be used to treat dysmenorrhoea and female infertility. The ground, roasted seeds can be used as a coffee substitute.

Growing extensively in Zimbabwe, the *Bauhinia* species scrambles up trees and shrubs, covering them with masses of frilly, white flowers in summer.

KANAK CHAMPA

Latin name *Pterospermum acerifolium*
Photographs show Left: flowers with leaves, single flower heads. Opposite: flowers and leaves on a stem.
Collected Ewanrigg Botanical Garden, Harare, Zimbabwe, 2015
Other common names Maple-leafed bayur tree, Bayur tree, Dinner plate tree.
Family *Malvaceae*
Origin Bangladesh, Bhutan, India, Myanmar, Nepal, Thailand.
Description A large tree that can grow up to 30m tall.
Leaves Palmately arranged leaves vary in size and shape; mature leaves can be 20–35cm long. They can be orbicular or oblong with a cordate base, entire or variously lobed, and sometimes peltate. Green above, they are silvery or grey and tomentose on the underside with dentate margins.
Flowers Bears large, white, fragrant blooms – axillary, solitary, or in pairs. Petals are linear to oblong, somewhat obliquely cuneate, and slightly shorter than the sepals, which can be up 18cm long. Flowers begin as one long bud, then separate into five more slender sepals as they mature.
Fruit Produces oblong, five-angled, woody, brown tomentose capsules, 10–15cm long.
Nomenclature *Pterospermum* comes from the Greek words, *pteron* meaning 'wing' and *sperma*, meaning 'seed'. *Acerifolium* describes the leaves, which are similar in shape to those of an *Acer*.
Uses Decorative ornamental and/or shade tree. The leaves, flowers, and wood have functions, from medicinal to ornamental. The common name 'dinner plate tree' comes from the fact that in India the huge leaves are put on moulds and formed into plates and bowls.

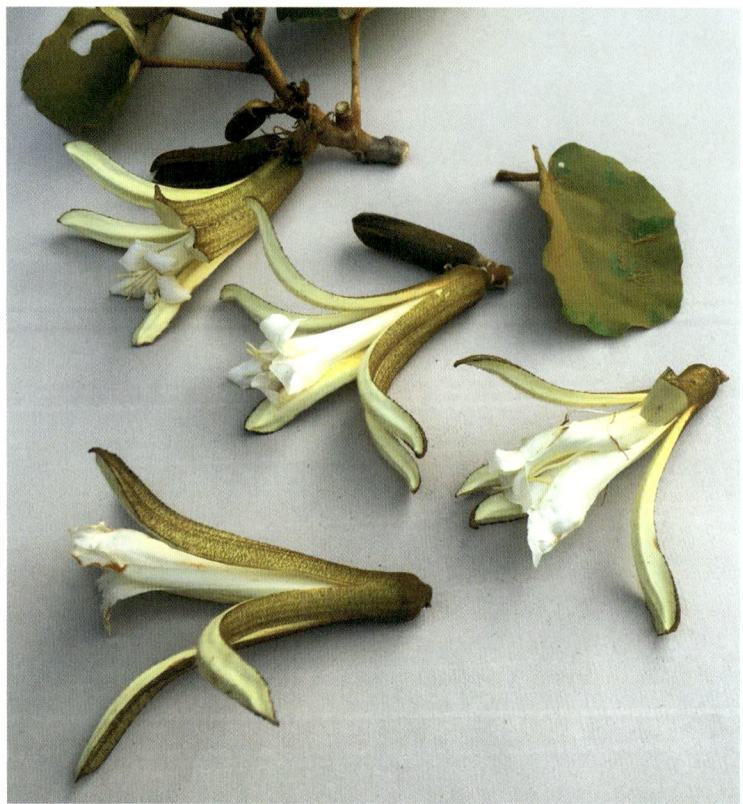

On visiting Ewanrigg, a botanical garden just outside Harare, I came across a huge specimen of this tree covered in flowers that resembled those of the vanilla orchid. Unable to identify it myself, I turned to the botanical community, and posted a picture of the tree on Instagram. Within a couple of hours I had an identification from the botanist Zaki Jamil in Malaysia, who confirmed that it was *Pterospermum acerfolium*. Zaki subsequently helped with much research for this book.

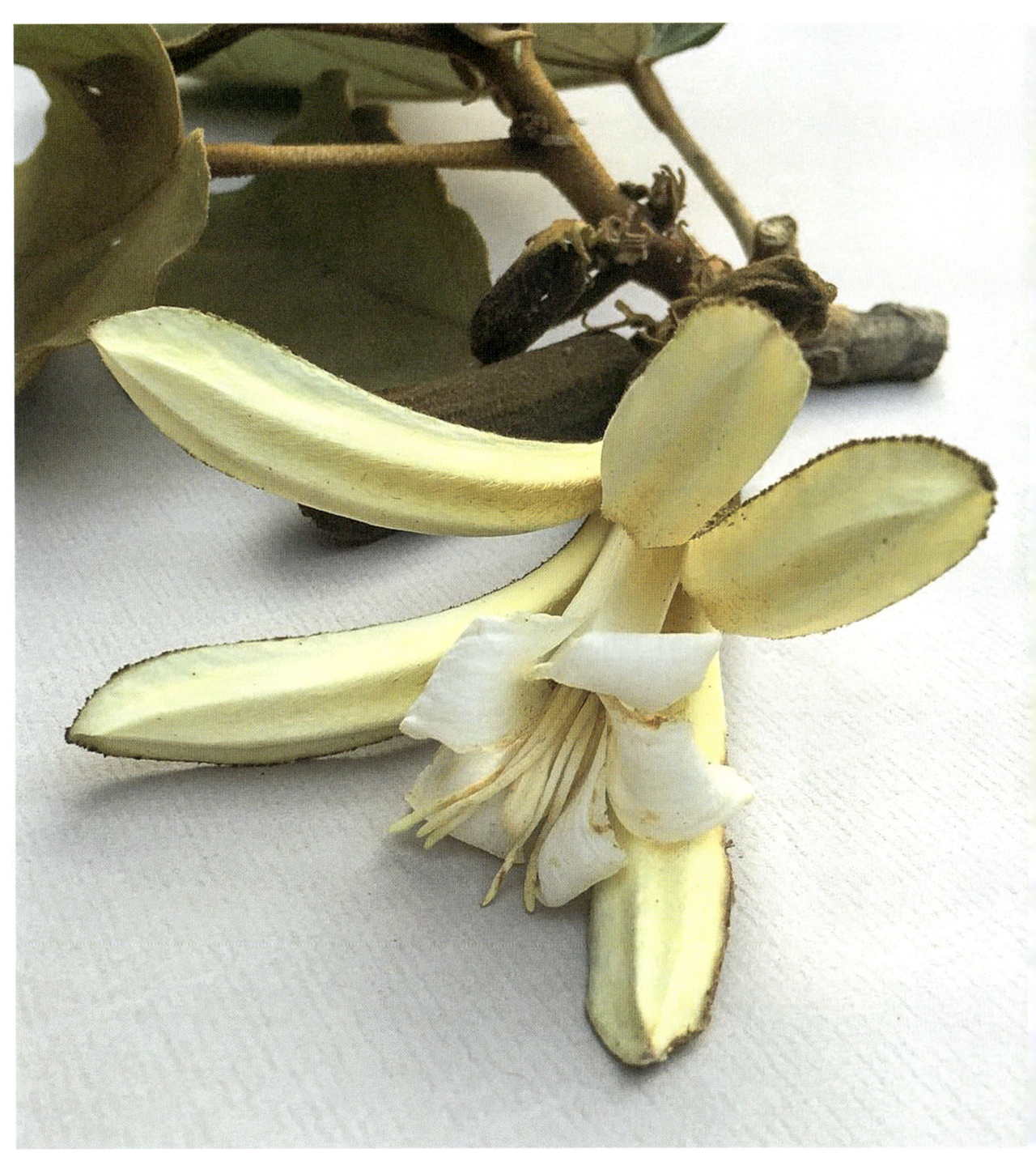

FIDDLE LEAF FRANGIPANI

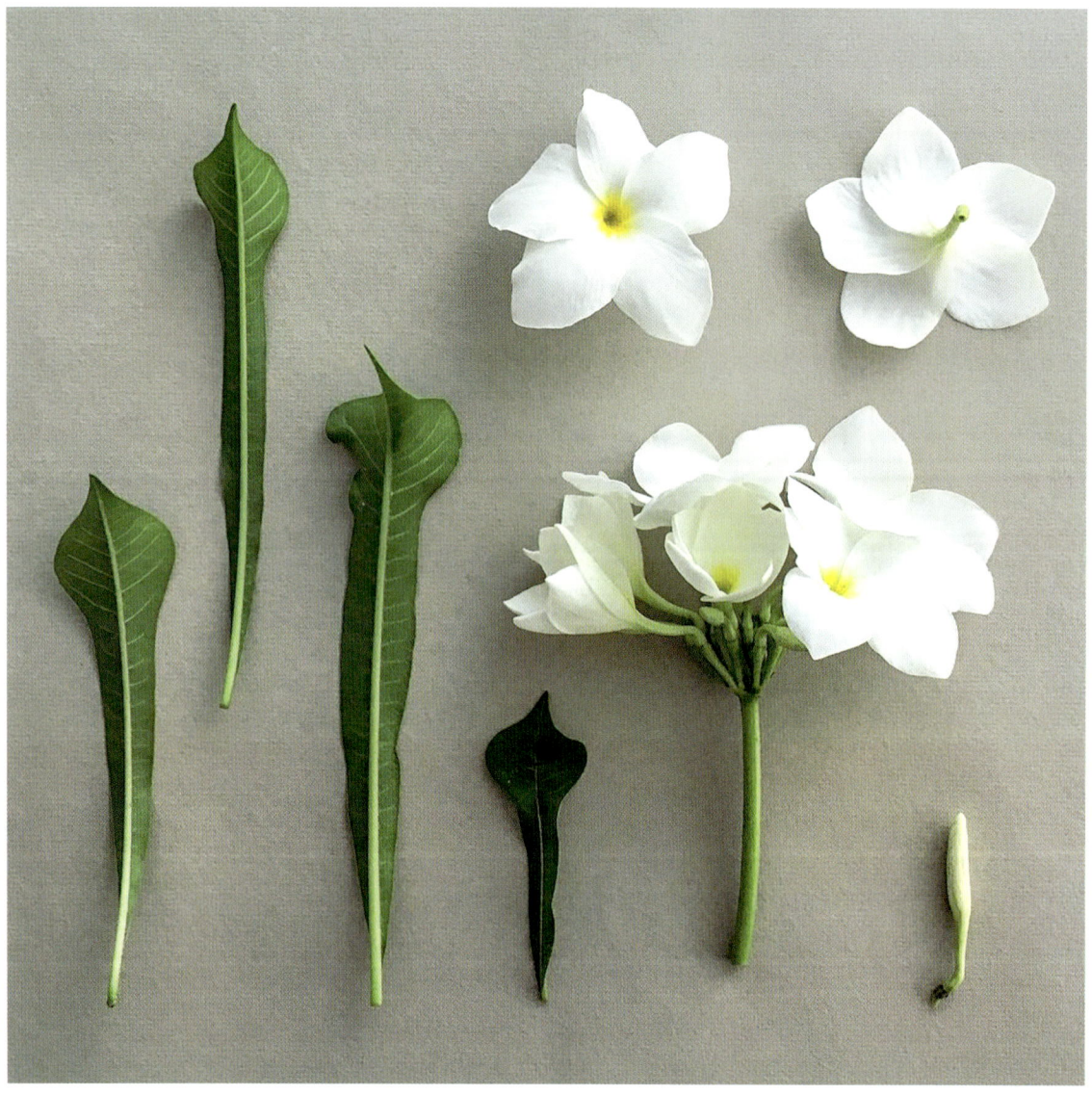

Latin name *Plumeria pudica*
Photograph shows Leaves, single flowers, cluster of flowers, bud.
Collected Pointe Aux Pimentes, Mauritius, 2017
Other common name Bridal Bouquet
Family *Apocynaceae*
Origin Tropical and subtropical Panama, Colombia, Venezuela.
Description An unusual, evergreen, ever-blooming shrub or tree that grows up to 4m tall.
Leaves Forms large, elongated, fiddle- or spoon-shaped, almost sessile, dark-green leaves without stalks.
Flowers Bears white flowers with yellow throats that bloom in bunches arranged like a bouquet, hence one of its common names – Bridal bouquet.
Fruit Forms two-part follicles that contain winged seeds.
Nomenclature Genus is named after the 17th-century French botanist Charles Plumier (1646–1704). *Pudica* is from Latin, meaning 'shy' or 'bashful'.
Uses Decorative ornamental.

Unlike most frangipanis, the flowers of this species are disappointingly devoid of perfume.

MOONFLOWER

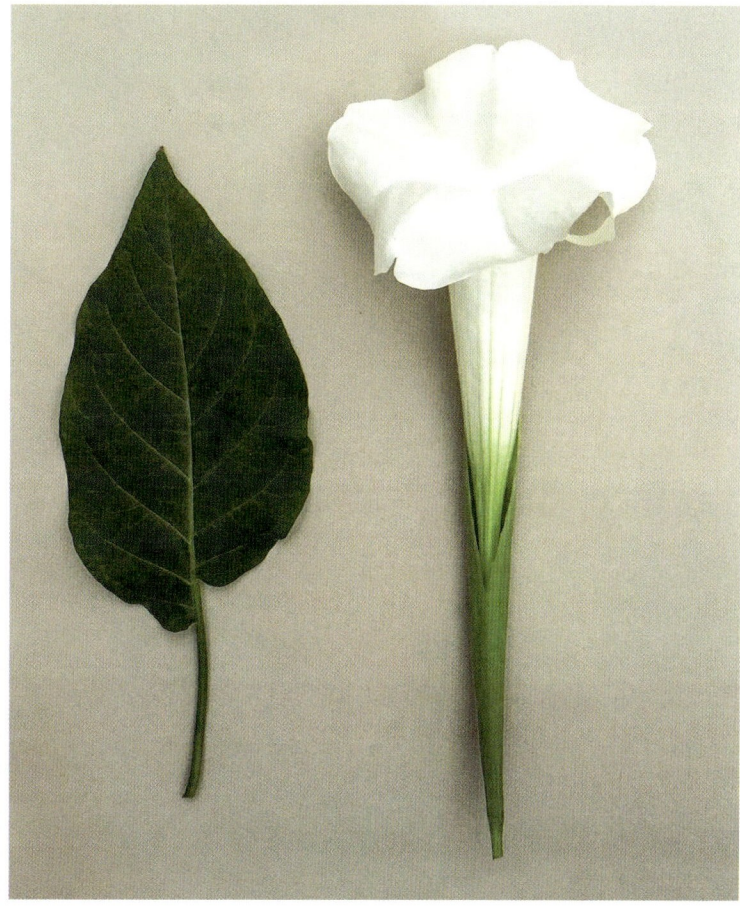

Ever present in the gardens of Maduma in shades of cream and pink, the *Brugmansia*'s huge flowers are showstoppers. They bloom in bursts following the cycle of the moon and have an intoxicating scent, which is most prominent at night. Growing easily in London's microclimate, I collected this from one of my own specimens.

Latin name *Brugmansia suaveolens*
Photograph shows Leaf, flower.
Collected London, UK, 2014
Other common names Angel's trumpet, Angel's tears
Family *Solanaceae*
Origin Coastal rainforests of south-eastern Brazil.
Description Semi-woody, evergreen shrub or small tree, growing up to 3–5m tall.
Leaves Elliptic, entire leaves are up to 25cm long and 15cm wide. Forms larger leaves when grown in shade.
Flowers Large, sweetly scented, trumpet-shaped flowers, usually white (but may be yellow or pink), are about 30cm long. The flowers are pendulous, hanging almost straight down from the stems.
Fruit Produces short, spindle-shaped capsules that change from green to brown when ripe.
Nomenclature The genus *Brugmansia* is named after the Dutch botanist Sebald Justinus Brugmans (1763–1819). *Saveolens* is from the Latin *suāveolenti*, meaning 'fragrant' or 'sweet smelling'.
Uses All parts of the plant are poisonous. Some indigenous South American groups use this plant as a psychoactive substance in their religious or spiritual ceremonies. It is also used as an analgesic in traditional medicine.

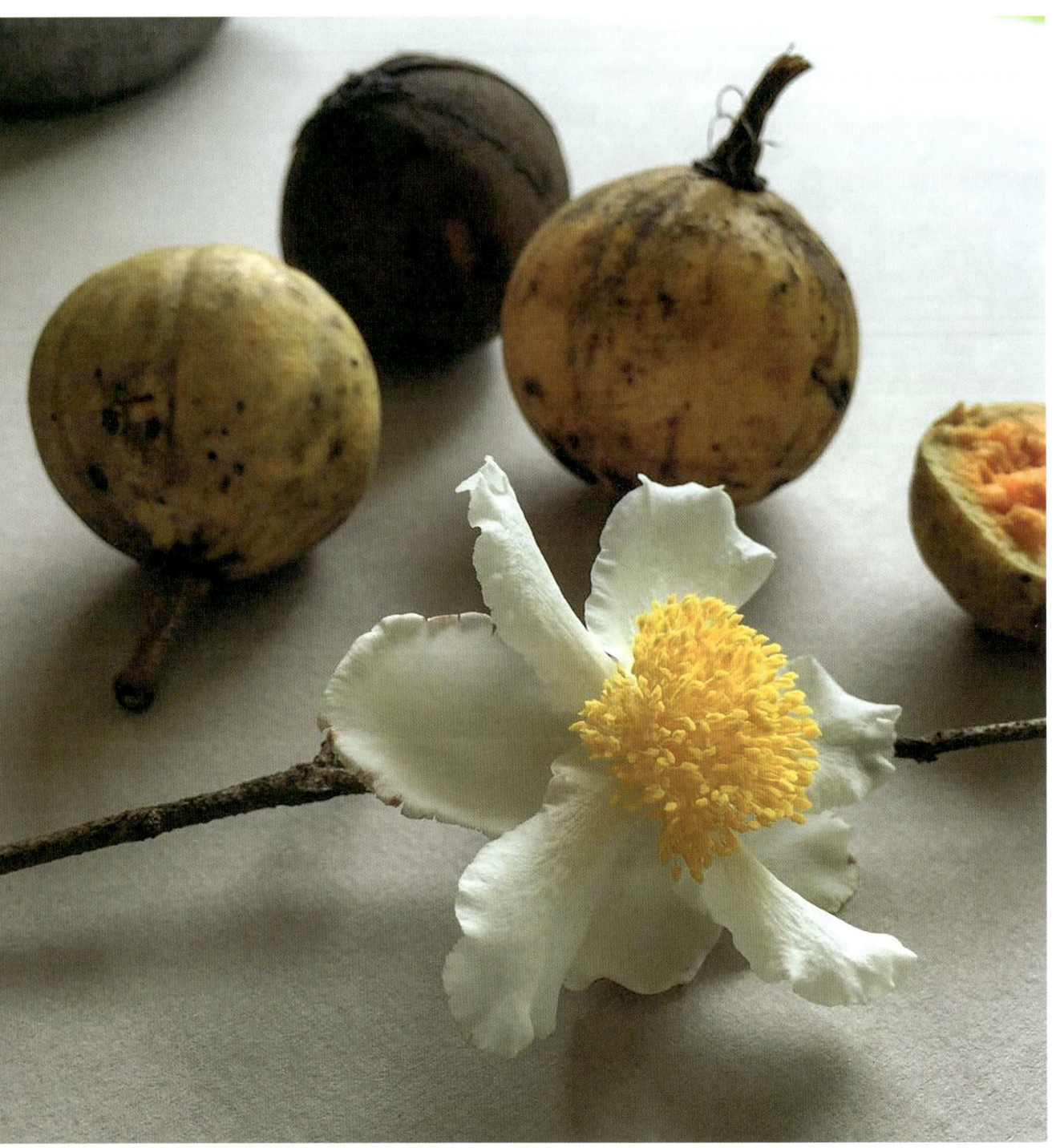

SNUFF BOX TREE

Latin name *Oncoba spinosa*
Photographs show Right: Flower, fruit, stem, leaves with bud. Opposite: fruit in various stages of maturity, open fruit, flower on a branchlet.
Collected Harare, Zimbabwe, 2016
Other common names Fried-egg tree, Fried-egg flower.
Family *Salicaceae*
Origin Tropical Africa – Senegal to Somalia, south to northern Angola, Zambia, Zimbabwe, South Africa.
Description A slow-growing, summer-flowering, evergreen, spiny tree that grows up to 5m tall.
Leaves Simple, shiny, bright-green leaves have paler undersides and serrated margins. They are 8–12cm long and 4–6cm wide.
Flowers Bears large, showy, sweetly scented, white flowers that measure 9cm across and have masses of yellow overlapping stamens in the centre.
Fruit Forms spherical, dark-green to brown, smooth-textured fruit, up to 6cm in diameter.
Nomenclature *Oncoba* is from *Onkub*, the Arabic name for the North African species, and *spinosa* refers to it spines.
Uses The dried fruit are used as rattles on percussion anklets worn by African dancers and hollowed-out to make snuff boxes, hence its common name.

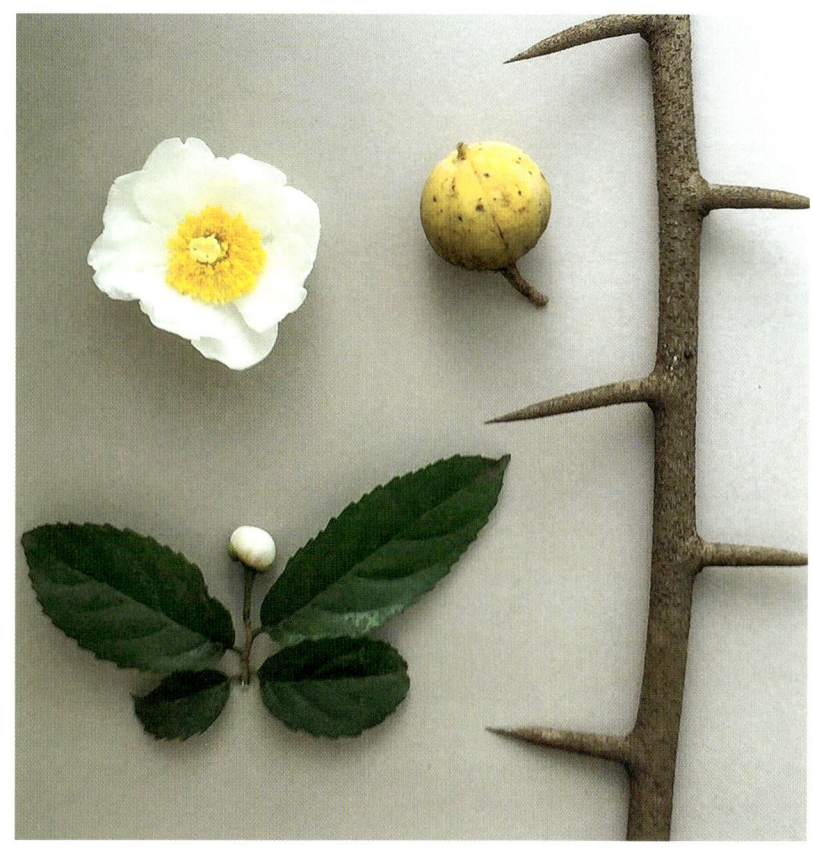

Lunching in a cafe in Harare I spotted a tree that I hadn't seen before flowering in the garden. With the owner's permission, I collected some plant material so that I could both photograph and identify it.

Achenes Small, dry, one-seeded fruit that does not open to release the seed.

Actinomorphic flowers Those that are 'star-shaped', or radial, and rotate from the central part, and can be divided into three or more identical sectors, that are related to each other (for example, a daisy).

Annual Plant that completes its lifecycle (germination to flowering) in the same year, then dies.

Anther (of a flower) Oval structure at the tip of a stamen that contains the pollen, or male gametophyte, essential for plant reproduction.

Aril Extra covering around a seed, typically coloured and fleshy or hairy.

Biennial Flowering plant that completes its lifecycle (germination to flowering) over two years. In the first year it forms leaves, stems, and roots, then lies dormant, and flowers in the second year.

Bifoliate Plants that have two leaves or pairs of leaves.

Bilobed (or bilobate) Consisting of, or having, two lobes.

Bulb A short, modified underground stem, or structure, made up of rings, or 'scales', that stores a plant's life cycle. Normally perennial.

Bulbous A plant that grows from a bulb.

Calyx (pl. Calices) Collective term for the sepals of a plant.

Carpel The ovary, style, and stigma of a plant. *See also* Pistil.

Corm A short, vertical, swollen underground stem that serves as a storage organ for some plants so they can survive adverse conditions, such as cold in winter or a summer drought

Corona Latin, meaning 'crown', this is the crown-, trumpet-, or funnel-shaped centre of some flower types, for example, daffodil or Lenten rose.

Cuneate plants Those that are wedge-shaped, with the narrow part at the bottom, or that have straight, or almost straight sides that meet at the base.

Cyme A flower cluster with a central stem that bears a single terminal flower which develops first; other flowers in the cluster develop as terminal buds of side stems.

Dehiscent capsule One that splits along a built-in line of weakness at maturity in order to release its contents.

Dentate (leaf margins) Toothed edges, or margins, to a leaf.

Domatia (singular *domatium*) Small chambers inside a plant or formed by hairs on leaves that to shelter arthropods.

Ellipsoid A three-dimensional shape (often capsule or fruit) that is elliptical through the long axis.

Epiphytic A plant that grows on the surface of another plant, taking its moisture and nutrients from the air, rain, water, and/or debris around it.

Hermaphrodite plant One that has both male and female organs. Also described as bisexual or monoecious.

Imparipinnate Plants with pinnately compound leaves in which there is a lone terminal leaflet rather than a terminal pair; also called 'odd-pinnate'.

Indumentum Covering of fine hairs on leaves, fruit, or capsules of a plant.

Inflorescence A group or cluster of flowers arranged on a stem. It may be one main branch or a complicated arrangement of branches.

Lanceolate (leaves) Leaf that is shaped like a spear, or lance, and tapers to a point at the tip (apex) or base; sometimes described as linear-lanceolate.

Lithophytes Plants that grow on rocks and derive moisture and nutrients from the air, rain, water, and/or debris around them; also known as lithophytic.

Loculicidal capsules Those that split longitudinally between the cavities or partitions, known as locules.

Margin Edge of a leaf.

Meristem Plant tissue, found chiefly at the growing tips of roots and shoots, which contains actively dividing cells that form new tissue.

Midrib (of a leaf) Central or principle vein of a leaf, which extends from the stem to its tip.

Modified stems *see Rhizomes*

Monotypic Having only one type, representative, or species of a genus.

Palmately arranged Leaves in which separate lobes or leaflets spread radially from a central point.

Panicle A loose, branching cluster of flowers – blooms arranged this way are said to be paniculate and are often not significant.

Pedate Leaves with radiating lobes that have deep clefts or are divided.

Pedicles Short floral stalks.

Peduncle The, usually green, stalk that supports an inflorescence of flowers.

Perennial Term used to describe a plant that lives for more than two years, but has no woody growth (i.e. is not a shrub or tree).

Perianth The outer parts of a flower.

Petiole Stalk that attaches a leaf to a plant stem.

Pistil The female part of a flower, comprising the stigma, style, and ovary.

Pseudo-bulb A bulb-like enlargement of the stem of orchids (especially epiphytic ones), where moisture is stored.

Pseudo-stem An apparent stem formed by overlapping leaf sheaths, that can enclose a 'true' stem, for example, in a banana plant.

Raceme Unbranched, indeterminate type of inflorescence that bears flowers on short floral stalks (pedicles) along its axis.

Rhizome A continuously growing modified, horizontal, underground stem that puts out lateral, or side shoots, and adventitious roots at intervals.

Scandent plants Those with a climbing habit whose stems grow upwards by attaching themselves to extraneous support often by means of runners.

Sepal The parts of the calyx of a flower, that encloses the petals – they are typically green and leaf-like.

Serrate Saw-like margins, or edges.

Spadix Spike of minute flowers closely arranged round a fleshy axis and typically enclosed in a spathe.

Stamen The male reproductive part of a part, comprising a filament with an anther at its tip.

Stigma The receptive tip of a carpel, or of several fused carpels. It is the part of a flower that receives pollen from pollinators such as bees, or moths.

Stipules Outgrowths (spines, glands, hairs, leaves) borne on either side (or sometimes just one side) of the base of a leafstalk.

Tepal A segment of the outer parts of a flower in which there is no differentiation between petals and sepals, for example lilies and tulips.

Tomentose plant One that has a leaf or plant capsule that is covered with densely matted woolly hairs.

Trifoliate Plants that have three leaves, or leaves that form in groups of three.

Trilobed (or trilobate) Consisting of, or having, three lobes.

Umbels An indeterminate inflorescence that consists of a number of short flower stalks spreading from a common point, like an umbrella.

INDEX OF PLANT NAMES AND PLANT FAMILIES

Abyssinian coral tree 24
Acanthaceae 20–21, 85
Acer 150
Aerangis fastuosa 128
African flame tree 41
African iris 50–51
African tulip tree 40
Alexandrian laurel 126
Aloe 36–37
Aloe *cameronnii* var. *bondana Reynolds* 36–37
Amaryllidaceae 118, 119–21, 124–25, 148
Ampolla tree 114–15
Anemone 130
Anemone coronaria 130
Angel orchid 129
Angel wing jasmine 136
Angel's tears 153
Angel's trumpet 153
Ansellia africana 54
Apocynaceae 56–57, 152
Apostle plant 70–71
Araceae 138–39
Arctic poppy 46–47
Aristolochia albida 91
Aristolochia clematitis 91
Aristolochia littoralis 90–91
Aristolochiaceae 90–91
Arundina gramanifolia 98
Asphodelaceae 36–37
Asteraceae 69

Bachelor's button 69
Ball tree 126
Bamboo orchid 98
Banksia rose 89
Bauhinia tree 110–11
Bauhinia fassoglense (syn.) 52
Bauhinia petersiana 149
Bauhinia x blakeana 110–11
Bayur tree 150–51
Bear's foot 131
Bignoniaceae 29, 40, 41, 43, 113
Birthwort 91
Black-eyed Susan 48
Blue bottle 69
Bombacaceae 62–63, 112, 114–15
Boraginaceae 66, 67
Bottle tree 62–63
Brachystegia spiciformis 42–43
Bridal bouquet 152
Brugmansia suaveolens 153
Brunfelsia grandiflora 77
Bull Bay magnolia 147

Caesalpinioiceae 27
Calico flower 90–91
Callophylum innophylum 126

Camellia 133
Camellia sinensis 133
Cameron's aloe 38–39
Cape coast lily 124–25
Cape gardenia 140–41
Cape leadwort 68
Cape plumbago 68
Carrion lily 102–103
Ceiba insignis 62–63
Ceiba speciosa 112
Centaurus cyanus 69
Chalice vine 55
Chequered daffodil 78
Chinaberry 92
Chinese forget me not 67
Chinese hibiscus 16–17
Chinese magnolia 104–105, 146
Chinese wisteria 86–87
Choriosa speciosa (syn.) 112
Christmas rose 131
Clematis florida 'Sieboldii' 82–83
Climbing lily 22–23
Clusiaceae 126
Coast coral tree 108–109
Coelogyne cristata 129
Colchicaceae 22–23
Common rothmannia 140–41
Common iris 72–75
Convolvulaceae 53
Copa de oro 55
Cornflower 69
Cosmos bipinnatus 134
Creeping bauhinia 52
Crinum macowanii 124–25
Crown imperial 34–35
Cup of gold 55
Cynoglossum amabile 67

Daffodil Mount hood 118
Delicious monster 138–39
Delonix regia 27
Dinner plate tree 150–51
Dietes bicolor 50–51
Drunken tree 62–63
Dutch iris 122–23
Dutch lily 38

Elegant Dutchman's pipe 90–91
Empress tree 93
Erythrina abyssinica 24
Erythrina caffra 108–109
Erythrina lysistemon 25
Ethiopian mahogany tree 60–61

Fabaceae 24, 25, 42–43, 86–87, 96–97, 108–109, 110–11, 149
Fernandoa magnifica 41
Fiddle leaf frangipani 152

Firewheel tree 26
Flamboyant tree 27
Flame lily 22–23
Flame tree 27, 40
Flaming parrot tulip 18–19
Flower of an hour 48
Flowering cherry 143
Flowering tobacco 132
Floss silk tree 112
Forest bell bush 85
Forest gardenia 142
Forget me not 66
Fortnight lily 50–51
Fountain tree 40
Foxglove tree 93
Frangipani 56–57
Fried-egg flower 154–55
Fried-egg tree 154–55
Fritillaria imperialis 35
Fritillaria mileagris 78
Fruit salad plant 138–39

Garden forget-me-not 66
Gardenia 142
Gardenia carinata 58–59
Gardenia posoquerioides 142
Gardenia thunbergia 142
German iris 72–75
Giant toad flower 102–103
Gloriosa lily 22–23
Gloriosa superba 22–23
Glory lily 22–23
Golden gardenia 58–59
Grandadilla 135
Great white cherry 143
Guinea hen flower 78–79
Gwangwadiza 52

Handroanthus impetiginosus 113
Harlequin flower 32–35
Hawaiian hibiscus 16–17
Helleborus niger 131
Helleborus orientalis 80–81
Hibiscus rosa-sinensis 16–17
Hibiscus trionum 48
Hong Kong orchid tree 110–11
Hymenocallis littoralis 148

Iceland poppy 46–47
Indian clock vine 20–21
Indian laurel 126
Imperial fritillary 34–35
Ipomoea obscura 53
Iridaceae 15, 32, 50–51, 70–71, 72–75
Iris germanica 72–75
Iris tingitana 122
Iris xiphium var. *praecox* 122
Iris x hollandica 122–23

Jasmine mango 56–57
Jasminum nitidum 136
Jasminum polyanthum 137
Jebicot 76

Kaiser's crown 34–35
Kalahari bauhinia 149
Kanak champa 150–51
Kapok tree 112
Kedah gardenia 58–59
Kigelia africana 29
Kiss me quick 77

Lady finger banana 28
Lady's slipper vine 20–21
Lance-leaved lily 38
Lathyrus odoratus 96–97
Lavender mist vanda 84
Lenten rose 80–81
Leopard orchid 54
Liberty iris 72– 75
Liliacea 18, 34–35, 38, 78–79
Lilium formosanum 134
Lilium tigrinum (syn.) 38
Loebner magnolia 144–45
Lotus 100–101
Lucky bean tree 25

Mackaya bella 85
Madras lilac 88–89
Mafura butter tree 60–61
Magnificent aerangis 128
Magnolia denudata 104
Magnolia grandiflora 147
Magnolia kobus 145
Magnolia liliflora 104
Magnolia stellata 145
Magnolia x loebneri 'Merrill' 144–45
Magnolia x soulangeana 104–107
Magnolia x soulangeana 'Lennei Alba' 146
Magnoliaceae 40, 104 –107, 144–45, 146, 147
Malvaceae 16–17, 48, 62–63, 112, 114–15, 150–51
Manchurian violet 76
Maple leaf bayur tree 150–51
Marama bean 52
Melia azedarach 92
Meliaceae 92
Mexican shell flower 14–15
Michelia champaca (syn.) 40
Monstera deliciosa 138–39
Moon flower 153
Moreae 51
Msasa tree 42–43
Musa acuminata 28
Musa cavendishii (syn.) 28

Musaceae 28
Myosotis sylvatica 66
Mysore clock vine 20–21

Narcissus 'Mount hood' 118
Narcissus poeticus 121
Narcissus 'Thalia' 119–21
Natal mahogany 60–61
Nelumbo nucifera 100–101
Nelumbonaceae 100–101
Neomarcia gracillis 70–71
Nicotiana sylvestris 132
Night-scented tobacco 132
Northeastern violet 76
Nuphar lutea 49
Nuphar lutea 'rubra' 99
Nymphaea lutea (syn.) 49, 99
Nymphaeaceae 49, 99

Obscure morning glory 53
Oleaceae 136, 137
Oncoba spinosa 154–55
Orchidaceae 54, 84, 98, 127, 128, 129

Papaver nudicaule 46–47
Papaveraceae 46–47
Paphiopedilum niveum 127
Passiflora edulis 135
Passiflora ligularis 135
Passiflora subpeltata 134
Passifloraceae 134, 135
Passion flower clematis 82–83
Passion fruit 135
Paulownia tomentosa 93
Paulowniaceae 93
Peacock flower 14–15
Persian lilac 92
Petrea volubilis 88–89
Pink coral tree 108–109
Pink jasmine 137
Pink lapacho tree 113
Pink trumpet tree 113
Plumbaginaceae 68
Plumbago auriculata 68
Plumbago capensis (syn.) 68
Plumeria alba 56–57
Plumeria pudica 152
Poet's daffodil 121
Pome banana tree 28
Princess tree 93
Proteaceae 26
Prunus 'Tai-haku'
Pseudobombax ellipticum 114–15
Pterospermum acerifolium 150
Purple wreath 88–89

Queen's wreath 88–89
Queensland firewheel tree 26

Ranunculaceae 80–81, 82–83, 130, 131
Red hot poker tree 24
Red water lily 99
Rhodes bush 68
Rongthao nari dok khao satun 127
Rosaceae 143
Rothmannia capensis 140–41
River bell 85
River crinum lily 124–25
River lily 124–25
Rose of China 16–17
Royal poinciana tree 27
Royal purple brunfelsia 77
Rubiaceae 58–59, 140–41, 142

Sacred coral tree 25
Sacred lotus 100–101
Salicaceae 154–55
Sandpaper vine 88–89
Saucer magnolia 104–107, 146
Sausage tree 29
Scrambling Dutchman's pipe 90–91
Shaving brush tree 114–15
Shining jasmine 136
Silk floss tree (pink) 112
Silk floss tree (yellow) 62–63
Snake tree 41
Snake's head fritillary 78–79
Snow white slipper orchid 127
Snow queen 129
Snuff box tree 154–55
Solandra maxima 55
Solonaceae 55, 77, 132, 153
Southern magnolia 147
Sparaxis pillansii 32–35
Spathodea campanulata 40
Spider lily 148
Split-leaf philodendron 138–39
Staghorn fern 56
Stapelia gigantea 102–103
Star jasmine 136
Star magnolia 145
Starfish flower 102–103
Stenocarpus sinuatus 26
Sweet pea 96–97
Swiss cheese plant 138–39

Tabebuia palmeri (syn.) 113
Tamanu 126
Tea plant 133
Tea shrub 133
Theaceae 133
Thunbergia mysorensis 20–21
Tiger flower 14–15
Tigridia pavonia 14–15
Transvaal kaffir boom 25
Tree gardenia 142
Triandrus daffodil 119–121

Trichilia emetica sub sp. *emetica* 60–61
Trichilia emetica sub sp. *superosa* 60–61
Trumpet daffodil 118
Tulip flower 26
Tulip magnolia 104–107
Tulip magnolia 'Lennei Alba' 146
Tulipa 'Flaming Parrot' 18–19
Tylosema fassoglensis 52

Vanda kanchana 'Lavender mist' 84
Vanda orchid 84
Venice mallow 48
Verbena bonariensis 134
Verbenaceae 88–89
Vlei lily 124 –25
Viola mandshurica 76
Violaceae 76

Walking iris 70–71
Water lily (red) 99
Water lily (yellow) 49
West Indian jasmine 56–57
White beefwood 26
White cedar 92
White dragon 62–63
White kapok 62–63
White oak 26
White passion flower 134
White silky oak 26
Wild gardenia 141
Wild iris 50–51
Windflower 130
Winter jasmine 137
Wisteria sinensis 86–87
Wood forget-me-not 66

Yesterday-today-and-tomorrow 77

Zebrawood tree 42–43
Zulu giant 102–103

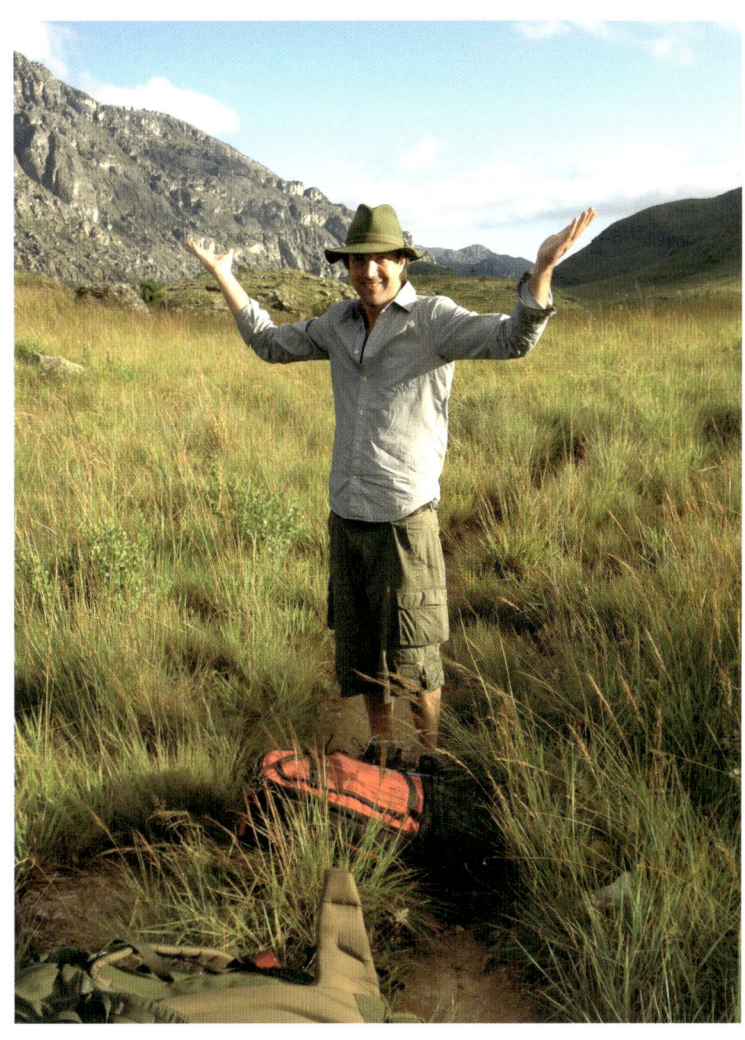

On sabbatical, Chimanimani, January 2012.

ACKNOWLEDGEMENTS

I feel an enormous sense of gratitude to my parents who recognised and fostered my love of flowers. They supported me in what was an unlikely pastime for a young boy growing up in Zimbabwe. Furthermore, they generously allowed me to study different disciplines which reflect in the various chapters and facets of my career. They have joined me on many collecting trips in Zimbabwe, and I have enjoyed their company immensely.

In acknowledgement of the people who influenced and affirmed me in my love of flora, I owe a debt of gratitude to:

Jessamine Brent, my great-grandmother, to whom flowers gave great joy. She didn't know any of the botanical names of the plants she grew, but the pleasure they gave her was evident. She said that she never felt closer to God than when she was in her garden, and I too feel close to the Divine when in nature.

Armenell Dupont, a notable plantswoman. I learned much at the knee of this rather fearsome woman who generously imparted her knowledge, and she remains one of my botanical heroes.

Bernie Cragg As part of our senior school curriculum we spent a week at Ringwood a camp up in the Nyanga mountains. The camp was run by Cragg, who recognised my knowledge and interest in the indigenous flora of the Eastern Highlands. He affirmed this attribute in a significant way that led to a new respect from my peers.

I would also like to thank all those who assisted in any way with the production of this book, in particular.

I am truly grateful to Catharine Snow and Simonne Waud for allowing me to share my photographs in hard copy.

To my editor Jemima Dunne for her conscientious and informed approach.

To book designer Bernard Higton, for his elegant layouts and patience as I navigated my way through my first book.

Zaki Jamil for undertaking the botanical research. Acquainted through Instagram, @zqqyjml is a talented botanist (the real deal) and a skilled botanical photographer who lives in Singapore. We have never met in person – until then Z.

My father, Scot Honey who undertook the project management of the images, copy, and botanical content of this book. He catalogued everything, creating spreadsheets to monitor progress and activity, and patiently paired up more than 250 images with botanical references and narratives and recalibrated my edits. This was all done with unfailing good humour, while cheering me on towards my deadline.

My mother, Toni Honey for her eye for detail and her valuable objective feedback on my compositions, when I struggled to resolve a layout. Writing in Harare, I worked nine hours a day, and my mama fed and watered me, cooking my favourite dishes, and also cheered me on to the finish line. Thanks, mama – our time picking sweet peas each morning was precious. This book is dedicated to you. I just hope you are still around to see it in print, if not catch you on the other side, I know you will be there.

Rosebie Morton, thanks for encouraging a fledgling florist and for kicking me out the nest.

Kimball Groves both a muse and amusing, you remain so.

Lulu Goodman Thank you for the years we created together.

Richard and Emma Honey, for your love, time, and support in running my floristry business.

Josh Smith, for the set of coloured pencils you gave me in 2007 with the note 'these are for your book' – one that I had not yet even conceived of at the time.

Will Johnson for the sketch book, on which this series was founded.

Jennifer Sturrock for kindling my creativity and journeying with me creatively as I navigated a new path.

Shane Connolly, our mutual admiration society continues. Thank you for your encouragement and friendship.

Katrina Lawson Johnston, KLJ, it's always a joy collaborating with you.

Twig Hutchinson from Minford Journal for allowing us to shoot the chapter openers in her house.

To all the Instagrammers who liked and encouraged me in my work and cheered me on, thanks, guys.

John Mccallen A shout out to @mccallen from Bellingham in Washington, another accidental botanist who is always able to give me a positive plant ID when I am unable to. Thanks John.